Status and Understanding of Groundwater Quality in the Two Southern San Joaquin Valley Study Units, 2005–2006: California GAMA Priority Basin Project

By Carmen A. Burton, Jennifer L. Shelton, and Kenneth Belitz

A product of the California Groundwater Ambient Monitoring and Assessment (GAMA) Program

Prepared in cooperation with the California State Water Resources Control Board

Scientific Investigations Report 2011–5218

U.S. Department of the Interior
U.S. Geological Survey

U.S. Department of the Interior
KEN SALAZAR, Secretary

U.S. Geological Survey
Marcia K. McNutt, Director

U.S. Geological Survey, Reston, Virginia: 2012

For more information on the USGS—the Federal source for science about the Earth, its natural and living resources, natural hazards, and the environment, visit http://www.usgs.gov or call 1–888–ASK–USGS.

For an overview of USGS information products, including maps, imagery, and publications, visit http://www.usgs.gov/pubprod

To order this and other USGS information products, visit http://store.usgs.gov

Suggested citation:
Burton, C.A., Shelton, J.L., and Belitz, Kenneth, 2012, Status and understanding of groundwater quality in the two southern San Joaquin Valley study units, 2005–2006—California GAMA Priority Basin Project: U.S. Geological Survey Scientific Investigations Report 2011–5218, 150 p.

Contents

Contents—Continued

Figures

Figures—Continued

Figures—Continued

Tables

Tables—Continued

Conversion Factors, Datums, and Abbreviations and Acronyms

Conversion Factors

Inch/foot/mile to SI

Multiply	By	To obtain
Length		
inch (in.)	2.54	centimeter (cm)
inch (in.)	25.4	millimeter (mm)
foot (ft)	0.3048	meter (m)
mile (mi)	1.609	kilometer (km)
Area		
square foot (ft^2)	0.09290	square meter (m^2)
square mile (mi^2)	2.590	square kilometer (km^2)
Radioactivity		
picocurie per liter (pCi/L)	0.037	becquerel per liter

Temperature in degrees Celsius (°C) may be converted to degrees Fahrenheit (°F) as follows:

$$°F = (1.8 \times °C) + 32$$

Datums

Vertical coordinate information is referenced to the North American Vertical Datum of 1988 (NAVD 88).

Horizontal coordinate information is referenced to the North American Datum of 1983 (NAD 83).

Specific conductance is given in microsiemens per centimeter at 25 degrees Celsius (µS/cm at 25 °C).

Concentrations of chemical constituents in water are given either in milligrams per liter (mg/L) or micrograms per liter (µg/L).

Abbreviations and Acronyms

AB	Assembly Bill (through the California State Assembly)
AL	action level
AL-US	U.S. Environmental Protection Agency action level
GAMA	Groundwater Ambient Monitoring and Assessment Program
HAL	lifetime health advisory level
HAL-US	U.S. Environmental Protection Agency lifetime health advisory level
HBSL	health-based screening levels
Hwy99T	Highway 99 understanding well
KERN	Kern County subbasin study unit
KING	Kings study area
KINGFP	Kings study area understanding well
KWH	Kaweah study area

Conversion Factors, Datums, and Abbreviations and Acronyms—Continued

Abbreviations and Acronyms—Continued

LRL	laboratory reporting level
LSD	land-surface datum
LUFT	leaking underground fuel tanks
MCL	maximum contaminant level
MCL-CA	California Department of Public Health maximum contaminant level
MCL-US	U.S. Environmental Protection Agency maximum contaminant level
MDL	method detection limit
NL	notification level
NL-CA	California Department of Public Health notification level
QC	quality control
RPD	relative percent difference
RSD5-US	U.S. Environmental Protection Agency risk-specific dose at a risk factor of 10^{-5}
SC	specific conductance
SESJ	Southeast San Joaquin Valley study unit
SMCL	secondary maximum contaminant level
SMCL-CA	California Department of Public Health secondary maximum contaminant level
SMCL-US	U.S. Environmental Protection Agency secondary maximum contaminant level
TEAP	terminal electron acceptor process
TLR	Tulare Lake study area
TT-US	USEPA treatment technique
TULE	Tule study area

Organizations

CDPH	California Department of Public Health (Department of Health Services prior to July 1, 2007)
CDPR	California Department of Pesticide Regulation
CDWR	California Department of Water Resources
LLNL	Lawrence Livermore National Laboratory
NAWQA	National Water-Quality Assessment Program (USGS)
SWRCB	State Water Resources Control Board (California)
USEPA	U.S. Environmental Protection Agency
USGS	U.S. Geological Survey

Conversion Factors, Datums, and Abbreviations and Acronyms—Continued

Selected chemical names

Ammonia	ammonia as nitrogen
$CaCO_3$	calcium carbonate
DBCP	1,2-dibromo-3-chloropropane
DO	dissolved oxygen
EDB	1,2-dibromoethane (ethylene dibromide)
NDMA	N-nitrosodimethylamine
Nitrate	nitrate as nitrogen
Nitrite	nitrite as nitrogen
PCE	tetrachloroethene (perchloroethene)
1,2,3-TCP	1,2,3-trichloropropane
TCE	trichloroethene
TDS	total dissolved solids
THM	trihalomethane
VOC	volatile organic compound

Units of Measure

cm	centimeter
$\delta^i E$	delta notation; the ratio of the heavier isotope (i) to the more common lighter isotope of an element (E), relative to a standard reference material, expressed as per mil
ft	foot (feet)
in.	inch
km	kilometer
km^2	square kilometer
L	liter
m	meter
mg/L	milligrams per liter (parts per million)
mi	mile
mi^2	square mile
µg/L	micrograms per liter (parts per billion)
µS/cm	microsiemens per centimeter
pCi/L	picocuries per liter
per mil	parts per thousand
pmc	percent modern carbon
TU	tritium unit

Status and Understanding of Groundwater Quality in the Two Southern San Joaquin Valley Study Units, 2005–2006: California GAMA Priority Basin Project

By Carmen A. Burton, Jennifer L. Shelton, and Kenneth Belitz

Abstract

Groundwater quality in the southern San Joaquin Valley was investigated from October 2005 through March 2006 as part of the Priority Basin Project of the Groundwater Ambient Monitoring and Assessment (GAMA) Program. The GAMA Priority Basin Project is conducted by the U.S. Geological Survey (USGS) in collaboration with the California State Water Resources Control Board and the Lawrence Livermore National Laboratory. There are two study units located in the southern San Joaquin Valley: the Southeast San Joaquin Valley (SESJ) study unit and the Kern County Subbasin (KERN) study unit.

The GAMA Priority Basin Project in the SESJ and KERN study units was designed to provide a statistically unbiased, spatially distributed assessment of untreated groundwater quality within the primary aquifers. The *status assessment* is based on water-quality and ancillary data collected in 2005 and 2006 by the USGS from 130 wells on a spatially distributed grid, and water-quality data from the California Department of Public Health (CDPH) database. Data was collected from an additional 19 wells for the *understanding assessment*. The aquifer systems (hereinafter referred to as primary aquifers) were defined as that part of the aquifer corresponding to the perforation interval of wells listed in the CDPH database for the SESJ and KERN study units.

The *status assessment* of groundwater quality used data from samples analyzed for anthropogenic constituents such as volatile organic compounds (VOCs) and pesticides, as well as naturally occurring inorganic constituents such as major ions and trace elements. The *status assessment* is intended to characterize the quality of untreated groundwater resources within the primary aquifers in the SESJ and KERN study units, not the quality of drinking water delivered to consumers.

Although the *status assessment* applies to untreated groundwater, Federal and California regulatory and non-regulatory water-quality benchmarks that apply to drinking water are used to provide context for the results.

Relative-concentrations (sample concentration divided by benchmark concentration) were used for evaluating groundwater. A relative-concentration greater than 1.0 indicates a concentration greater than the benchmark and is classified as high. The relative-concentration threshold for classifying inorganic constituents as moderate or low was 0.5; for organic constituents the threshold between moderate and low was 0.1.

Aquifer-scale proportion was used as the primary metric for assessing the quality of untreated groundwater for the study units. High aquifer-scale proportion is defined as the areal percentage of the primary aquifers with a high relative-concentration for a particular constituent or class of constituents. Moderate and low aquifer-scale proportions were defined as the areal percentage of the primary aquifers with moderate and low relative-concentrations, respectively. Two statistical approaches—grid-based and spatially weighted—were used to evaluate aquifer-scale proportions for individual constituents and classes of constituents. Grid-based and spatially weighted estimates were comparable for the two study units in the southern San Joaquin Valley (within 90 percent confidence intervals).

The *status assessment* showed that inorganic constituents were more prevalent than organic constituents and that relative-concentrations were higher for inorganic constituents than for organic constituents. For inorganic constituents with human-health benchmarks, the relative-concentration of at least one constituent in the SESJ study unit was high in 30 percent of the primary aquifers. In the KERN study unit, the relative-concentration of at least one constituent was high in 23 percent of the primary aquifers. In the SESJ and KERN study units, the inorganic constituents with human-health benchmarks detected at high relative-concentrations in more than 2 percent of the primary aquifers were arsenic, boron, vanadium, nitrate, uranium, and gross alpha radioactivity. Additional constituents with human-health benchmarks— antimony, radium, and fluoride—were detected at high relative-concentrations in the KERN study unit.

For inorganic constituents with aesthetic benchmarks (secondary maximum contaminant levels, SMCLs), the relative-concentration of at least one constituent in the SESJ study unit was high in 6.6 percent of the primary aquifers. In the KERN study unit, the relative-concentration of at least one constituent was high in 22 percent of the primary aquifers. Inorganic constituents with aesthetic benchmarks detected at high relative-concentrations in the primary aquifers in the SESJ and KERN study units were iron and manganese. Additional constituents with aesthetic benchmarks—total dissolved solids (TDS), sulfate, and chloride—were detected at high relative-concentrations in the KERN study unit.

In contrast, the *status assessment* for organic constituents with human-health benchmarks showed that relative-concentrations were high in 4.8 percent and 2.1 percent of the primary aquifers in the SESJ and KERN study units, respectively. The special-interest constituent, perchlorate, was detected at high relative-concentrations in 1.2 percent of the primary aquifers in the SESJ study unit.

Twenty-eight of the 78 VOCs (not including fumigants) analyzed were detected. Of these 28 VOCs, benzene had high relative-concentrations in the SESJ study unit, and relative-concentrations for the other 27 VOCs were moderate and low. Five of the 10 fumigants were detected; 1,2-dibromo-3-chloropropane (DBCP) was the only fumigant with high relative-concentrations in the SESJ and KERN study units.

Of the 136 pesticides and pesticide degradates analyzed, 33 were detected. Human-health benchmarks were established for eighteen of the detected pesticides. Dieldrin was detected at moderate relative-concentrations in the SESJ and KERN study units. All other pesticides detected with human-health benchmarks were present at low relative-concentrations. The detection frequencies for two of these pesticides—simazine and atrazine—were greater than or equal to 10 percent in the SESJ and KERN study units.

The *understanding assessment* of groundwater quality included an analysis of correlations of selected water-quality constituents or classes of constituents with potential explanatory factors. The understanding assessment indicated that the concentrations of many trace elements and major ions were correlated to well depth, groundwater age, and/or geochemical conditions. Many trace elements were positively correlated with depth. Arsenic, boron, vanadium, fluoride, manganese, and iron concentrations increased with well depth or depth to top-of-perforations. The concentrations for these trace elements also were higher in older (pre-modern) groundwater. In contrast, uranium concentrations decreased with increasing depth and groundwater age.

Most trace elements were correlated to geochemical conditions. Arsenic, antimony, boron, fluoride, manganese, and iron concentrations generally were higher wherever the pH of the groundwater was greater than 7.6. Concentrations for these constituents generally were higher at low concentrations of dissolved oxygen (DO). Uranium was the exception; uranium concentrations generally were lower at high pH and at high concentrations of DO.

Nitrate concentrations generally were lower in deeper wells. Nitrate concentrations also were higher in groundwater with higher DO.

Total dissolved solids, sulfate, and chloride concentrations were higher in the KERN study unit than in the SESJ study unit. Total dissolved solids were negatively correlated with pH in the KERN study unit. Total dissolved solids and sulfate were higher in areas with more agricultural land use. Chloride concentrations increased with depth to top-of-perforations in the KERN study unit.

Organic constituents and constituents of special interest, like many inorganic constituents, were correlated with well depth, groundwater age, and DO. Unlike most trace elements, however, solvent and pesticide detections, and total trihalomethanes (THM), DBCP, and perchlorate concentrations decreased with increasing well depth. Volatile organic compound, solvent, and pesticide detections, and THM concentrations also were lower in older (pre-modern) groundwater than in modern-age groundwater. Solvent detections and total THM, DBCP, and perchlorate concentrations increased with increasing DO concentrations.

Introduction

Groundwater composes nearly one-half of the water used for public supply in California (Hutson and others, 2004). To assess the quality of ambient groundwater in aquifers used for drinking-water supply and to establish a baseline groundwater quality monitoring program, the California State Water Resources Control Board (SWRCB), in collaboration with the U.S. Geological Survey (USGS) and Lawrence Livermore National Laboratory (LLNL), implemented the Groundwater Ambient Monitoring and Assessment (GAMA) Program (California State Water Resources Control Board, 2011, website at http://www.waterboards.ca.gov/water_issues/programs/gama). The statewide GAMA Program consists of three projects: (1) the Priority Basin Project, conducted by the USGS (U.S. Geological Survey, 2011, website at http://ca.water.usgs.gov/gama); (2) the Domestic Well Project, conducted by the SWRCB; and (3) the Special Studies, conducted by LLNL. On a statewide basis, the GAMA Priority Basin Project primarily focused on the deep part of the groundwater resource (primary aquifers), and the SWRCB Domestic Well Project generally focused on the shallow aquifer systems. The primary aquifers may be at less risk of contamination than the shallow wells, such as private domestic or environmental monitoring wells, that are closer to surficial sources of contaminants. As a result, concentrations of contaminants, such as volatile organic compounds (VOCs) and nitrate, in wells screened in the deep primary aquifers may be lower than concentrations of contaminants in shallow wells (Nolan and Hitt, 2006; Landon and others, 2010).

The SWRCB initiated the GAMA Program in 2000 in response to Legislative mandates (State of California, 1999, 2001a, Supplemental Report of the 1999 Budget Act 1999–00 Fiscal Year). The GAMA Priority Basins Project was initiated in response to the Groundwater Quality Monitoring Act of 2001 (State of California, 2001b, Section 10780-10782.3 of the California Water Code, Assembly Bill 599) to assess and monitor the quality of groundwater in California. The GAMA Priority Basin Project is a comprehensive assessment of statewide groundwater quality designed to improve understanding of and to identify risks to groundwater resources, and to increase the availability of information about groundwater quality to the public. For the GAMA Priority Basin Project, the USGS, in collaboration with the SWRCB, developed the monitoring plan to assess groundwater basins through direct and other statistically reliable sampling approaches (Belitz and others, 2003; California State Water Resources Control Board, 2003). Additional partners in the GAMA Priority Basin Project are the California Department of Public Health (CDPH), the California Department of Pesticide Regulation (CDPR), the California Department of Water Resources (CDWR), local water agencies, and well owners (Kulongoski and Belitz, 2004).

The range of hydrologic, geologic, and climatic conditions in California must be considered in an assessment of groundwater quality. Belitz and others (2003) partitioned the State into 10 hydrogeologic provinces, each with distinctive hydrologic, geologic, and climatic characteristics (fig. 1). These hydrogeologic provinces include groundwater basins designated by the CDWR (California Department of Water Resources, 2003). Groundwater basins generally consist of relatively permeable, unconsolidated deposits of alluvial or volcanic origin (California Department of Water Resources, 2003). Eighty percent of California's approximately 16,000 active and standby drinking-water wells listed in the statewide database maintained by the CDPH (hereinafter referred to as CDPH wells) are located in designated groundwater basins within these hydrologic provinces. Some groundwater basins, such as the San Joaquin Valley basin, cover large areas and are further divided into groundwater subbasins by the CDWR. Groundwater basins and subbasins were prioritized for sampling on the basis of the number of CDPH wells in the basin or subbasin, with secondary consideration given to municipal groundwater use, agricultural pumping, the number of historical leaking underground fuel tanks, and registered pesticide applications (Belitz and others, 2003). Of the 472 basins and subbasins designated by the CDWR, 116 basins, as well as additional areas outside defined groundwater basins, were grouped into 35 study units, which include approximately 95 percent of CDPH wells in California.

The two GAMA Priority Basin Project study units (fig. 1) located in the southern San Joaquin Valley are the Southeast San Joaquin Valley study unit (hereinafter referred to as the SESJ study unit) and the Kern County Subbasin study unit (hereinafter referred to as the KERN study unit). The SESJ study unit is composed of four CDWR groundwater subbasins (Kings, Kaweah, Tule, and Tulare Lake, fig. 2). KERN is composed of one groundwater subbasin (Kern County, fig. 2).

Purpose and Scope

This report is one of a series of GAMA Priority Basin Project assessment reports presenting the *status* and *understanding* of current water-quality conditions in GAMA Priority Basin Project study units. Tabulated USGS data are available from several study units and are available as data series reports (for example, Burton and Belitz, 2008; Shelton and others, 2008), and planned subsequent reports will address changes or *trends* in water quality across time.

The *status* and *understanding assessments* of the two southern San Joaquin Valley study units are presented in this report. The purposes of this report are to provide (1) a *study unit description*: brief description of the hydrogeologic setting of the study units; (2) a *status assessment*: an assessment of the current status of untreated-groundwater quality in the primary aquifers; and (3) an *understanding assessment*: an identification of the natural and human factors affecting groundwater quality and an explanation of the relations between water quality and selected potential explanatory factors. An explanation of the causative factors of any relations between water quality and explanatory factors is beyond the scope of this report.

The *status assessment* in this report includes analysis of water-quality data from 130 wells selected for sampling by the USGS within spatially distributed grid cells across the two study units (hereinafter referred to as USGS-grid wells). Eighty-two percent of the USGS-grid wells were wells listed in the CDPH database, and the remainder had perforation intervals similar to CDPH wells in each study unit. Samples were collected from USGS-grid wells for analysis of anthropogenic constituents such as volatile organic compounds (VOCs) and pesticides, as well as naturally occurring constituents such as major ions, nutrients, and trace elements. Water-quality data from the CDPH database also were used to supplement data collected by the USGS for the GAMA Priority Basin Project. The resulting set of water-quality data from USGS-grid wells and supplemental CDPH data were considered to be representative of the primary aquifers in the southern San Joaquin Valley study units.

To provide context, the water-quality data discussed in this report were compared to California and Federal regulatory and non-regulatory benchmarks for drinking water. The assessments in this report are intended to characterize the quality of untreated groundwater resources of the primary aquifers in the study units, not the drinking water delivered to consumers by water purveyors. This study does not attempt to evaluate the quality of water delivered to consumers.

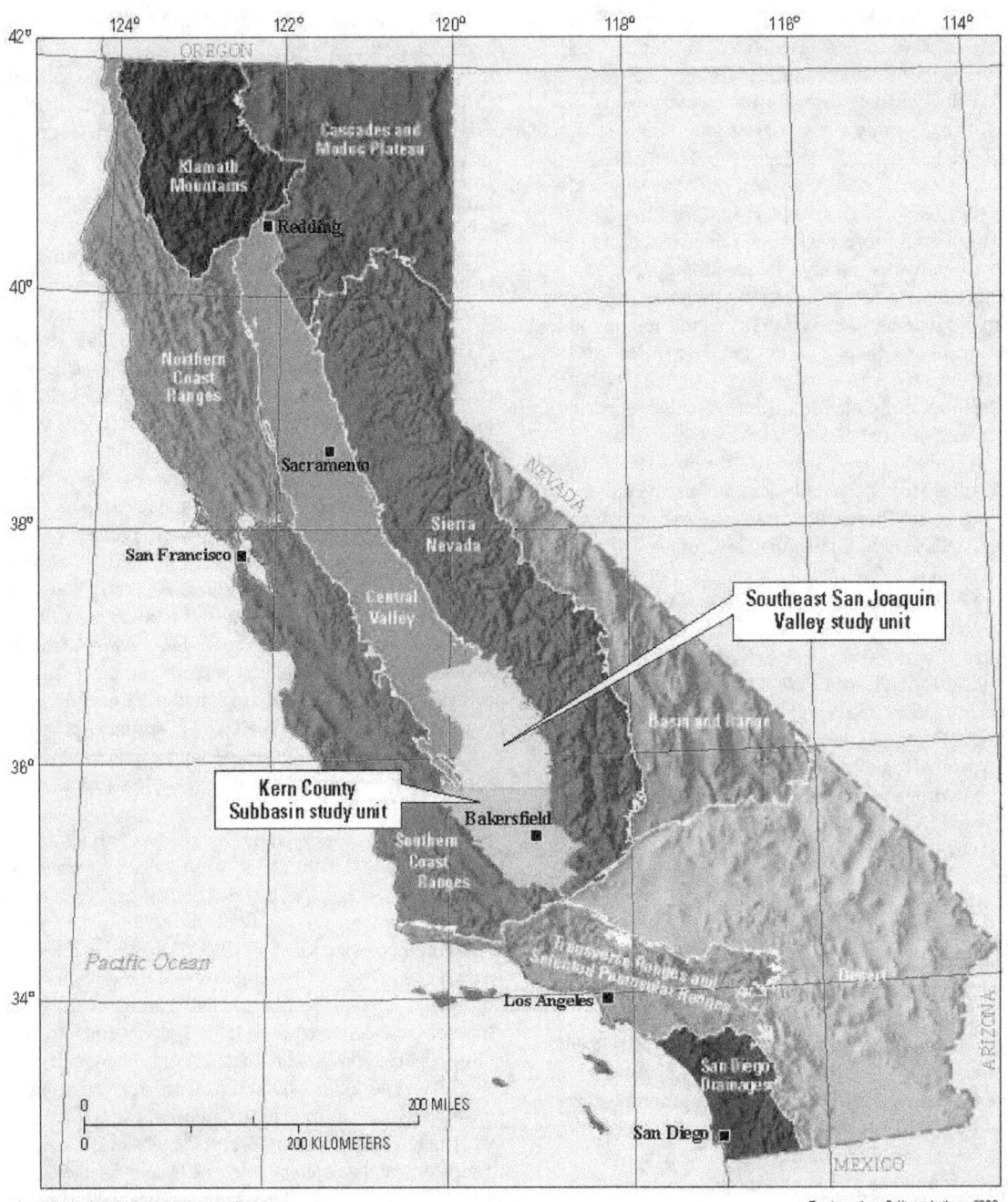

Shaded relief derived from U.S. Geological Survey
National Elevation Dataset, 2006,
Albers Equal Area Conic Projection

Provinces from Belitz and others, 2003

Figure 1. Location of the two southern San Joaquin Valley study units—Kern County Subbasin and Southeast San Joaquin Valley—and California hydrogeologic provinces (modified from Belitz and others, 2003), California GAMA Priority Basin Project.

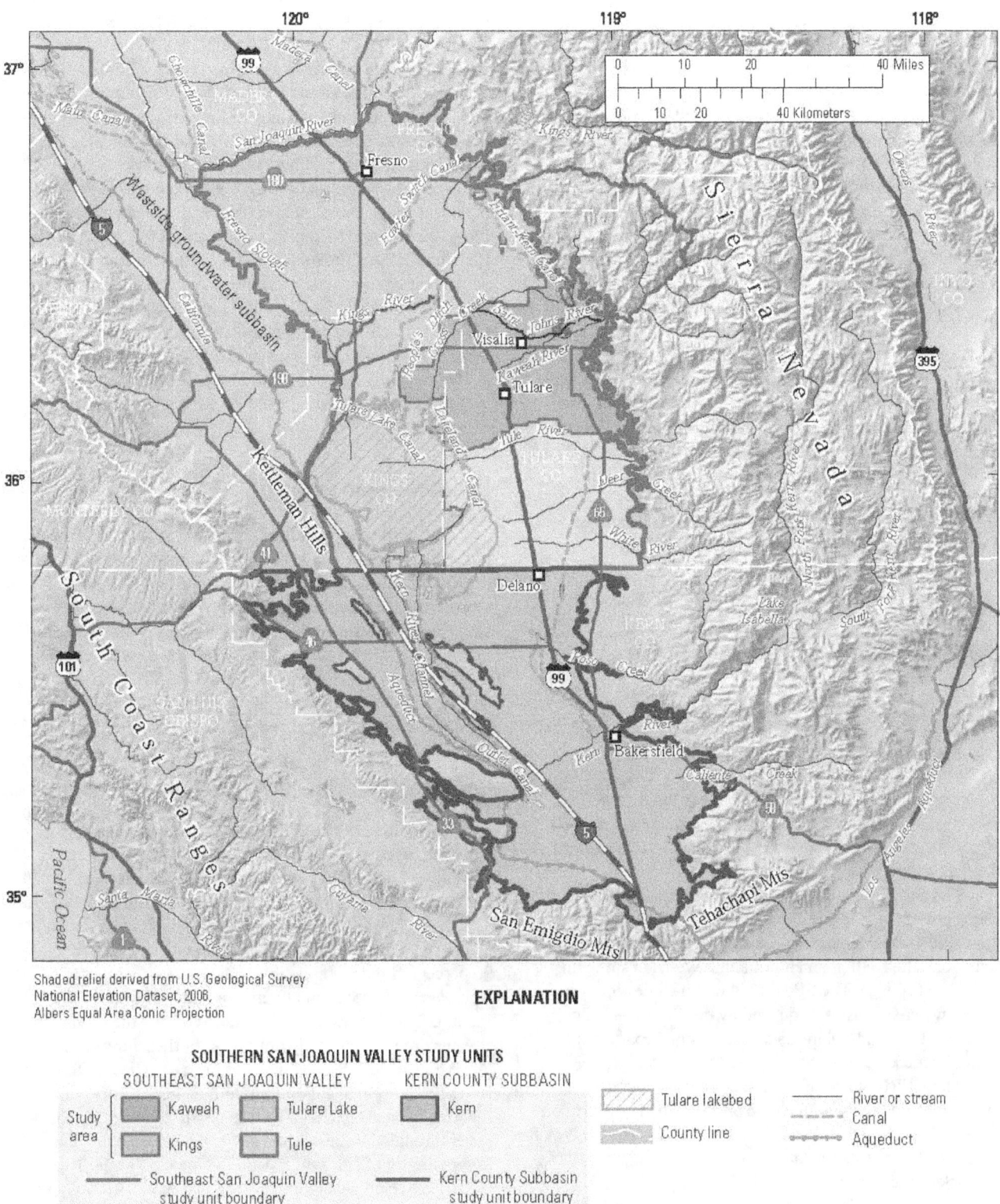

Shaded relief derived from U.S. Geological Survey
National Elevation Dataset, 2006,
Albers Equal Area Conic Projection

EXPLANATION

SOUTHERN SAN JOAQUIN VALLEY STUDY UNITS

SOUTHEAST SAN JOAQUIN VALLEY   KERN COUNTY SUBBASIN

Study area:
- Kaweah
- Tulare Lake
- Kern
- Kings
- Tule

- Tulare lakebed
- County line

- River or stream
- Canal
- Aqueduct

—— Southeast San Joaquin Valley study unit boundary

—— Kern County Subbasin study unit boundary

Figure 2.   Study areas and geographic features of the two southern San Joaquin Valley study units (Southeast San Joaquin Valley and Kern County Subbasin), California GAMA Priority Basin Project.

The *understanding assessment* for the SESJ and the KERN study units includes data from 19 wells sampled by USGS for the purpose of understanding (hereinafter referred to as USGS-understanding wells). Some of the USGS-understanding wells represent shallow zones above the primary aquifers. Potential explanatory factors examined included land-use, well-depth and perforation information, position in the groundwater flow system, groundwater age, density of septic systems and historically leaking underground fuel tanks, and geochemical-condition indicators. A comprehensive analysis of all possible explanatory factors is beyond the scope of this report.

Water-quality data for samples collected by the USGS for the GAMA Priority Basin Project in the SESJ and KERN study units and details of sample collection, analysis, and quality-assurance procedures were reported by Burton and Belitz (2008) and Shelton and others (2008). Using these data and data from the CDPH database, this report describes methods used in designing the sampling network, identifying CDPH data for use in the *status assessment*, estimating aquifer-scale proportions, analyzing ancillary datasets, classifying groundwater age, and assessing the status and understanding of groundwater quality and its relation to selected explanatory factors.

Description of the Study Units

The southern San Joaquin Valley lies in the Central Valley Hydrogeologic Province described by Belitz and others (2003) and includes two study units of the GAMA Priority Basin Project—the Southeast San Joaquin Valley (SESJ) and the Kern County Subbasin (KERN) study units. Together, these two study units cover about 6,780 square miles (mi²) [17,560 square kilometers (km²)] in Fresno, Kings, Tulare, and Kern Counties.

The SESJ study unit covers about 3,780 mi² (9,790 km²) and is bounded by the San Joaquin River on the north, the Sierra Nevada Mountains to the east, Kern County to the south, and the Kettleman Hills and the Westside subbasin to the west (fig. 2). The SESJ study unit includes four groundwater subbasins as defined by the CDWR—Kings, Kaweah, Tule, and Tulare Lake (California Department of Water Resources, 2003).

The KERN study unit is at the southern boundary of the Central Valley hydrogeologic province, and covers about 3,000 mi² (7,770 km²) in Kern County (fig. 1). The study unit is bounded by the Kern County line to the north, the granitic bedrock of the Sierra Nevada and Tehachapi Mountains to the east and southeast, and the marine sediments of the San Emigdio Mountains and South Coast Ranges to the

southwest and west (fig. 2) (California Department of Water Resources, 2006e). The KERN study unit is composed of one groundwater subbasin as defined by the CDWR—Kern County (California Department of Water Resources, 2003).

The land use in the southern San Joaquin Valley primarily is agricultural, with most of the irrigated acreage used for field crops and fruit and nut orchards. Land use in the SESJ study unit is 85 percent agricultural, 9 percent natural, and 6 percent urban based on the classification of USGS National Land Cover Data (Nakagaki and others, 2007; figs. 3 and 4). Land use in the KERN study unit is 66 percent agricultural, 31 percent natural, and 3 percent urban. Urban land use in the SESJ and KERN study units is relatively low; however, the rates of increase in population in Fresno and Bakersfield from 1990 to 2000, the largest cities in the SESJ and Kern study units, respectively, are the highest in the San Joaquin Valley and two of the highest in the State (U.S. Census Bureau, 2000).

Hydrogeologic Setting

The northwest-trending, asymmetrical structural trough of the San Joaquin Valley occupies the southern two-thirds of the Central Valley and is filled with marine and continental sediments of Tertiary and Quaternary age. The sediment thickness increases from the valley margins toward the axis of the trough, and from the north toward the south, up to a total thickness of about 30,000 feet (ft) in the south-central area of Kern County (Page, 1986). Freshwater occurs in the uppermost 3,000 ft, with brackish water beneath (Page, 1973).

Southeast San Joaquin Valley Study Unit

The climate for the SESJ study unit is Mediterranean with hot, dry summers and cool, moist winters. Average rainfall across the study unit ranges from 7 inches (in.) [18 centimeters (cm)] in the western part of the study unit to 13 in. (33 cm) in the eastern part of the study unit (California Department of Water Resources, 2006a, b, c, d). More than 90 percent of the precipitation falls between October and April. Several creeks and rivers drain the study unit. The San Joaquin, Kings, Kaweah, and Tule Rivers are the primary streams draining the study unit; most of their flow originates in the Sierra Nevada Mountains to the east. The San Joaquin River flows north and empties into the Sacramento–San Joaquin Delta. The Kings River is diverted to the San Joaquin River through the Fresno Slough. The Kaweah and Tule Rivers flow south or west toward the location of the Tulare Lake bed, which generally has been dry since 1919 as a result of long-term climate changes and stream diversions (Lofgren and Klausing, 1969; Faunt, 2009).

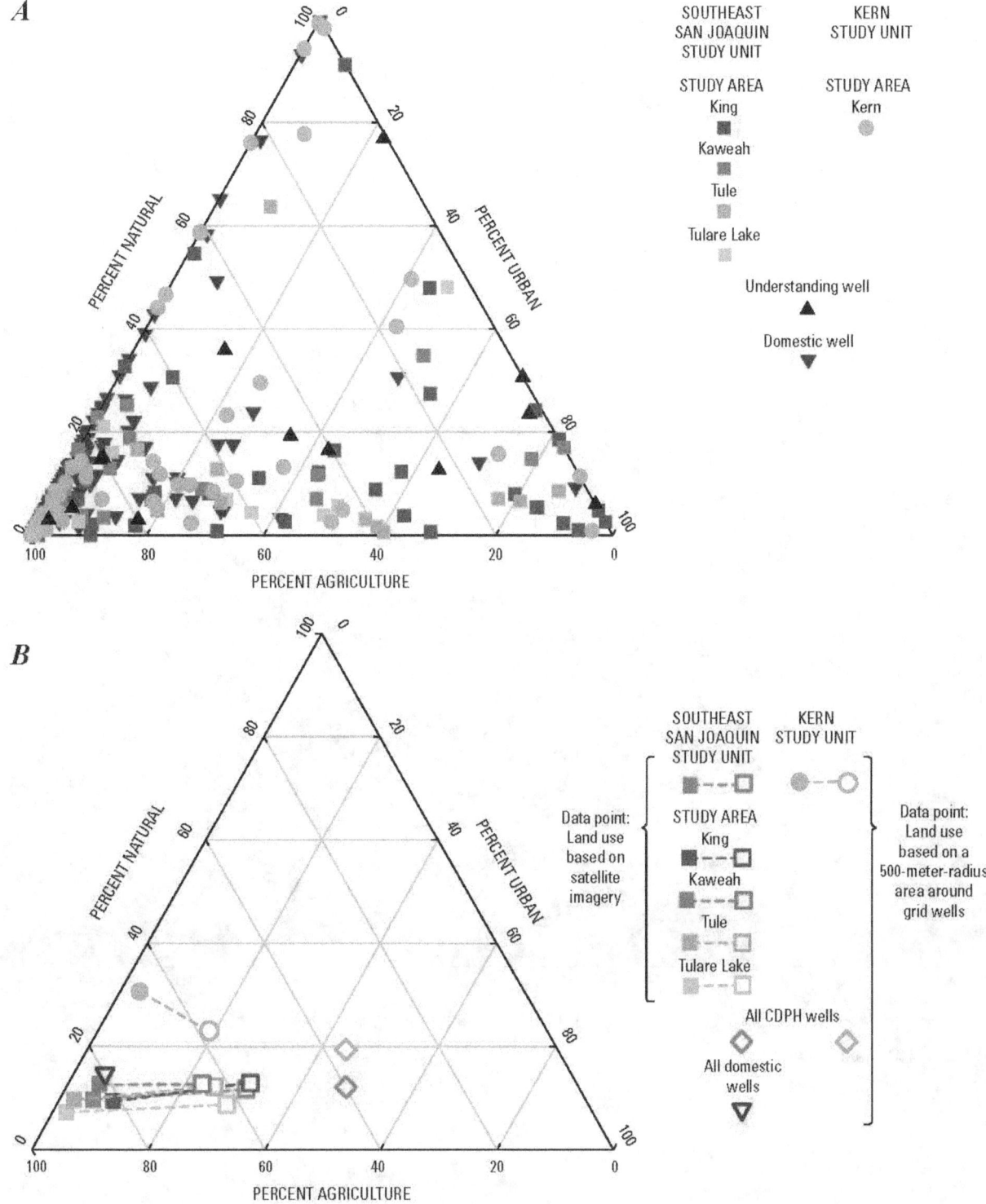

Figure 3. Ternary diagram with proportions of urban, agricultural, and natural land use for (A) wells and (B) study units, study areas, and domestic wells, Southeast San Joaquin Valley and Kern County Subbasin study units, California GAMA Priority Basin Project.

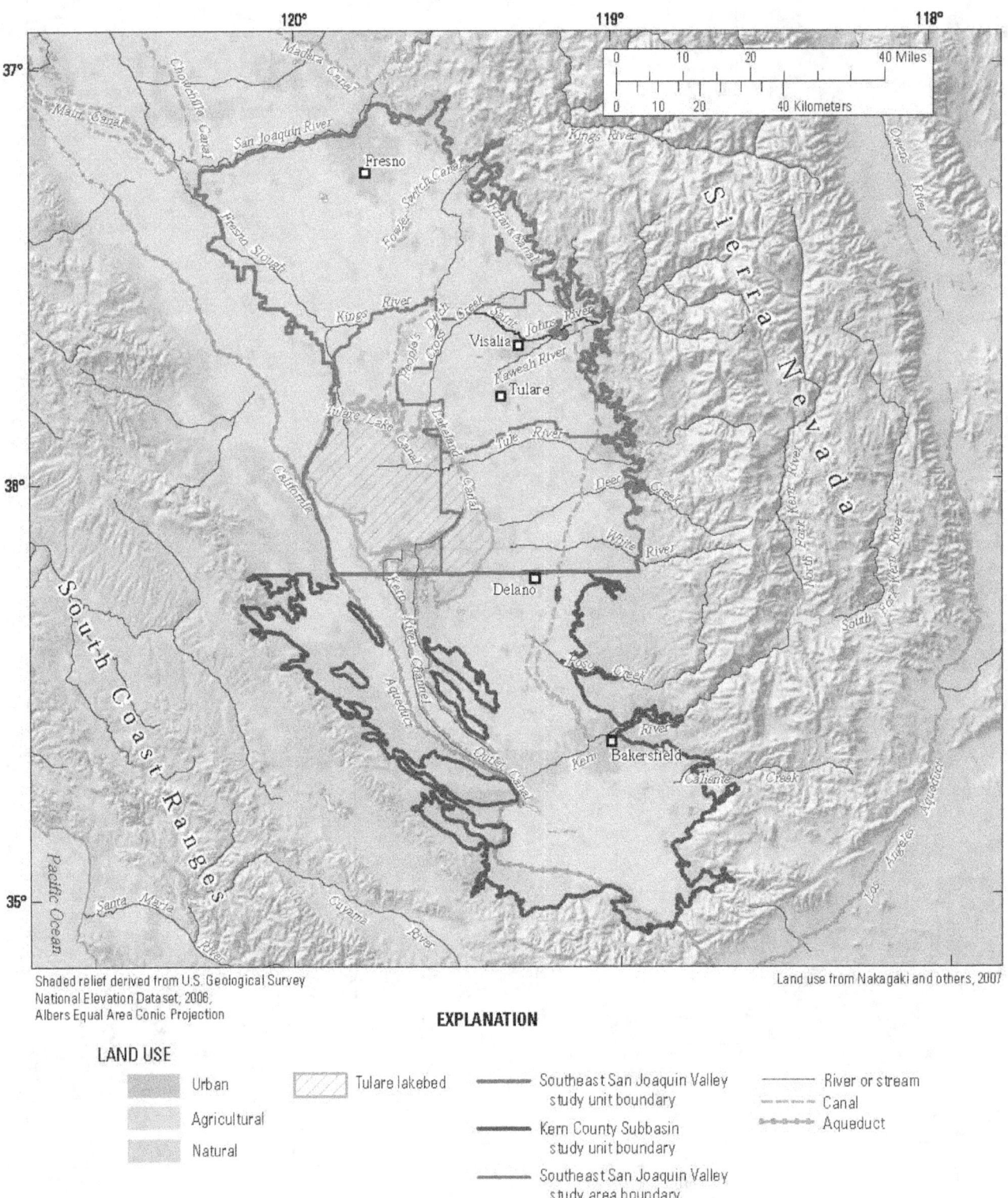

Shaded relief derived from U.S. Geological Survey
National Elevation Dataset, 2006,
Albers Equal Area Conic Projection

Land use from Nakagaki and others, 2007

EXPLANATION

LAND USE

Urban

Agricultural

Natural

Tulare lakebed

—— Southeast San Joaquin Valley study unit boundary

—— Kern County Subbasin study unit boundary

—— Southeast San Joaquin Valley study area boundary

—— River or stream

— — Canal

●—●—● Aqueduct

Figure 4. Land use in the Southeast San Joaquin Valley and Kern County Subbasin study units, California GAMA Priority Basin Project.

The primary source of recharge is runoff from the nearby Sierra Nevada Mountains (Faunt, 2009). Other sources of recharge include irrigation, seepage from rivers, streams, and irrigation canals, percolation of precipitation, urban runoff, and recharge (California Department of Water Resources, 2003; Wright and others, 2004). Discharge from the aquifer primarily is from groundwater pumping for irrigation and public water supply. About 40 percent of the water used in the study unit comes from groundwater (California Department of Water Resources, 2003). Most municipalities in the study unit, such as Visalia and Tulare, use groundwater as their sole source of public supply (Water Education Foundation, 2006).

The SESJ study unit is divided into four study areas—the Kings, Kaweah, Tule, and Tulare Lake study areas—which coincide with the groundwater subbasins, of the same names, that were defined by the CDWR.

Kings Study Area

The Kings study area includes 1,530 mi² (3,960 km²) of Fresno County and the northern part of Kings and Tulare Counties. It is bounded to the north by the San Joaquin River, to the east by the Sierra Nevada Mountains, to the south by the Kaweah and Tulare Lake groundwater subbasins, and to the west by the Delta-Mendota and Westside groundwater subbasins. The San Joaquin and Kings Rivers are the two major rivers within or bordering the subbasin (California Department of Water Resources, 2006a) (fig. 2).

The aquifer system in the Kings study area consists of unconsolidated marine and continental deposits of the Tertiary and Quaternary age overlain by younger alluvial deposits of Quaternary age. The eastern two-thirds of the study area consist of the quaternary alluvial fans of the Kings River, with compound alluvial fans of intermittent streams to the north and south (Burow and others, 1997). These deposits generally are highly permeable. The western one-third of the study area, near the center of the valley, includes less permeable deposits in the basin geomorphic province (Davis and others, 1959; Faunt, 2009). Vertical flow is restricted by discontinuous silt and clay layers to an increasing degree toward the valley center. The Corcoran Clay member of the Tulare Formation, ranging in depths of about 250–550 ft (75–170 m), is the most extensive clay layer and forms a regional confining unit (California Department of Water Resources, 2006a) (fig. 5).

Kaweah Study Area

The Kaweah study area includes 700 mi² (1,810 km²) of Tulare County. It is bounded by the Kings subbasin to the north, the Sierra Nevada Mountains to the east, the Tule subbasin to the south, and the Tulare Lake subbasin to the west. The Kaweah and Saint Johns Rivers are the two major rivers within the subbasin. The primary sources of groundwater recharge are recharge from the landscape (mostly irrigation return flow) and the Kaweah River (California Department of Water Resources, 2006b; Faunt, 2009) (fig. 2).

The aquifer system in the Kaweah subbasin consists of unconsolidated marine and continental deposits of Pliocene, Pleistocene, and Holocene age (California Department of Water Resources, 2006b). The eastern part of the subbasin consists of three stratigraphic layers: continental deposits, older alluvium, and younger alluvium. The deeper continental deposits are from the Pliocene and Pleistocene age and are poorly permeable. The older alluvium, consisting of quaternary alluvial deposits, is moderately-to-highly permeable and is the major aquifer in the subbasin. Also, the younger alluvium, consisting of quaternary alluvial fan deposits, is moderately-to-highly permeable. The alluvial fan deposits interfinger with the less permeable deposits in the basin geomorphic province in the western part of the subbasin. The groundwater aquifer is confined by the Corcoran Clay in the western part of the subbasin (fig. 5) (California Department of Water Resources, 2006b; Faunt, 2009).

Tule Study Area

The Tule study area includes 730 mi² (1,890 km²) of Tulare County. The study area is bounded by the Kaweah subbasin to the north, the Sierra Nevada Mountains to the east, Kern County to the south, and the Tulare Lake subbasin and Kings County to the west. Tule River and White River are the major rivers within the subbasin, and they flow westward toward the Tulare Lake bed (California Department of Water Resources, 2006c) (fig. 2).

The aquifer system in the Tule subbasin consists of continental deposits of Pliocene, Pleistocene, and Holocene age (California Department of Water Resources, 2006c). The basin and lake deposits in the western part of the subbasin consist mainly of silt and clay and are relatively impermeable. The Corcoran Clay of the Tulare Formation in the western part of the subbasin (fig. 5) is a confining layer for underlying groundwater. The older alluvium underlies the alluvial fans, is very permeable, and is a major aquifer in the subbasin. The younger alluvial fans are very permeable but contain little water. The continental deposits in the eastern part of the subbasin are poorly sorted deposits of clay, silt, sand, and gravel from the Sierra Nevada Mountains and are a major source of groundwater (California Department of Water Resources, 2006c).

Tulare Lake Study Area

The Tulare Lake study area includes 820 mi² (2,120 km²) of Kings County. It is bounded by the Kings subbasin to the north, the Kaweah and Tule subbasins to the east, Kern County to the south, and the Kettleman Hills and Westside subbasin to the west. The southern part of the subbasin is located in the former Tulare Lake bed (California Department of Water Resources, 2006d) (fig. 2).

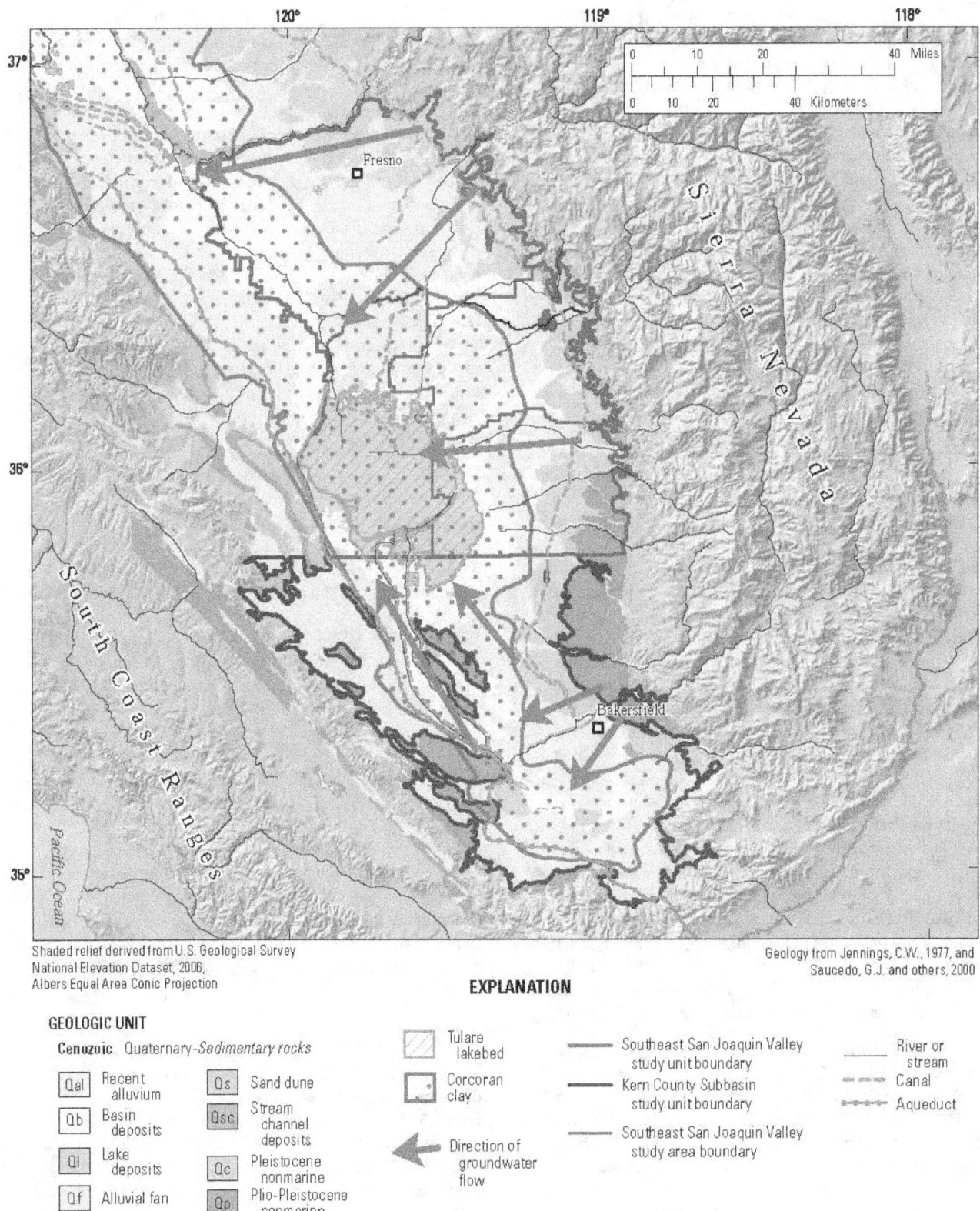

Figure 5.   Geology and generalized regional groundwater flow system in the two southern San Joaquin Valley study units, California GAMA Priority Basin Project.

The aquifer system of the Tulare Lake subbasin consists of older and younger alluvium, and continental deposits (California Department of Water Resources, 2006d). Continental deposits are poorly sorted deposits of clay, sand, and gravel; their permeability is low-to-moderate. Older alluvium is poorly sorted deposits of clay, sand, silt, and gravel. The older alluvium is moderately to highly permeable and is a major aquifer in the subbasin. The younger alluvium consists of sorted and unsorted beds of clay, sand, silt, and gravel. The younger alluvium is very permeable but lies above the water table. Flood-basin and lake deposits, which include the Corcoran Clay, are not an important source of groundwater in the subbasin (California Department of Water Resources, 2006d) (fig. 5).

In general, regional groundwater flow is from the east toward the west. North of the Kings River, groundwater generally flows westward toward the valley axis at the western boundary of the study unit. South of the Kings River, groundwater generally flows west or southwest toward the Tulare Lake bed (fig. 5).

Kern County Subbasin Study Unit

The hot Mediterranean climate of the KERN study unit is well suited to farming, with long, hot summer days and cool nights, and mild, damp winters with dense fog. The average annual precipitation, on the basis of a 56-year record at the California Irrigation Management Information System stations in Bakersfield and Shafter, located about 30 miles (78 km) west of Bakersfield, is 6 in. (15 cm), which occurs as rain primarily between November and February (California Irrigation Management Information System, 2006).

The Kern River is the primary stream draining into the study unit (fig. 2); however, the primary sources of recharge are from artificial recharge at groundwater banking facilities (Tom Haslebacher, Kern County Water Agency, written commun., June 15, 2007). Locations of groundwater banking facilities are given in Shelton and others (2008). Secondary sources of recharge include return flows from agricultural and municipal irrigation and infiltration of flows from intermittent streams along the margin of the subbasin (California Department of Water Resources, 2006e).

The primary geologic formations that compose the aquifer system are the Plio-Pleistocene-age Tulare and Kern River Formations and overlying alluvium and terrace deposits (California Department of Water Resources, 2006e). The aquifer system is unconfined in the eastern part and above the Corcoran Clay, where it is present, in the central and western parts of the study unit. Aquifers below the Corcoran Clay are confined (Dale and others, 1966). The aquifer is thickest along the eastern margin of the study unit and ranges from about 175 to 2,900 ft (320 to 880 m) with an average thickness of about 600 ft (183 m), according to estimates from the California Department of Water Resources (2006e).

In general, groundwater flows westward out of the Sierra Nevada and Tehachapi Mountains toward the center of the valley. Groundwater north of the Kern River flows west then flows north toward the Tulare Lakebed. Groundwater to the south of the Kern River flows toward southwest (fig. 5).

Methods

The *status assessment* provides a spatially unbiased assessment of groundwater quality in the primary aquifers, whereas the *understanding assessment* was designed to evaluate the natural and human factors that affect groundwater quality in the southern San Joaquin Valley study units. This section describes the methods used for: (1) defining groundwater quality, (2) assembling the datasets used for the status assessment, (3) determining which constituents warrant assessment, (4) calculating aquifer-scale proportions, and (5) analyzing statistics for the *understanding assessment*. Methods used for compilation of data on potential explanatory factors are described in appendix A.

The primary metric for defining groundwater quality is *relative-concentration*, which compares concentrations of constituents measured in groundwater to regulatory and non-regulatory benchmarks used to evaluate drinking-water quality. Constituents were selected for additional evaluation in the assessment based on objective criteria by using their relative-concentrations. Groundwater-quality data collected by GAMA Priority Basin Project and data compiled in the CDPH database are used in the *status assessment*. Two statistical methods based on spatially unbiased equal-area grids are used to calculate aquifer-scale proportions of low, moderate, or high relative-concentrations: the "grid-based" method uses one value per cell to represent groundwater quality, and the "spatially weighted" method uses many values per cell (Belitz and others, 2010).

Priority Basin Project *understanding assessments* are designed for the evaluation of the natural and human factors that affect groundwater quality at the study-unit level. The understanding assessments can be compared with other study units at regional and statewide scales. A finite set of potential explanatory factors was analyzed in relation to constituents of interest to place the observed water quality within the context of physical and chemical processes. Statistical tests were used to identify significant correlations between the constituents of interest and potential explanatory factors.

Status Assessment Methods

The *status assessment* included the following two steps. (1) Water-quality data were normalized to their respective water-quality benchmarks by calculating their relative-concentrations (Toccalino and others, 2004; Toccalino and Norman, 2006). (2) Aquifer-scale proportions were determined for categories of "high," "moderate," and "low" relative-concentrations using two methods: grid-based and spatially weighted. Results for the two approaches were compared, and results from the preferred approach were used to identify constituents of interest for further discussion.

Relative-Concentrations and Water-Quality Benchmarks

Concentrations of constituents are presented as relative-concentrations in the *status assessment* section of this report:

$$\text{Relative-concentration} = \frac{\text{Sample concentration}}{\text{Water-quality benchmark concentration}}.$$

Relative-concentrations provide context for the measured concentrations in the sample: relative-concentrations less than 1.0 indicate sample concentrations less than the benchmark, and values greater than 1.0 indicate sample concentrations greater than the benchmark. The use of relative-concentrations permits comparison of a wide range of concentrations for different constituents on a single scale.

Toccalino and others (2004), Toccalino and Norman (2006), and Rowe and others (2007) used the ratio of measured concentration to a benchmark (either maximum contaminant level [MCLs] or health-based screening levels [HBSL]) and defined this ratio as the benchmark quotient. Relative-concentrations used in this report are equivalent to the benchmark quotient reported by Toccalino and others (2004) for constituents with water-quality benchmarks. Relative-concentrations were computed only for compounds with water-quality benchmarks; therefore, constituents without water-quality benchmarks were not included in the status assessment. HBSLs were not used in this report because HBSLs are not currently used as benchmarks by California drinking-water regulatory agencies.

Regulatory and non-regulatory benchmarks apply to water that is served to the consumer, not to untreated groundwater. However, to provide some context for the water-quality results, concentrations of constituents measured in the untreated groundwater were compared with regulatory and non-regulatory health-based benchmarks established by the U.S. Environmental Protection Agency (USEPA) and CDPH (U.S. Environmental Protection Agency, 2006; California Department of Health Services, 2007). The health-based benchmarks used for each constituent were selected in the following order of priority:

1. Regulatory, in the order of health-based USEPA and CDPH maximum contaminant levels (MCL-US and MCL-CA), USEPA action levels (AL-US), and USEPA treatment technique levels (TT-US). Federal benchmarks were used unless the California levels were lower.

2. Non-regulatory USEPA and CDPH secondary maximum contaminant levels (SMCL-US and SMCL-CA). For constituents with recommended and upper SMCL-CA levels, the values for the upper levels were used.

3. Non-regulatory, in the order of health-based CDPH notification levels (NL-CA), USEPA lifetime health advisory levels (HAL-US) and USEPA risk-specific dose (1 in 100,000 lifetime risk of cancer, RSD5-US).

Note that for constituents with multiple types of benchmarks, this hierarchy may not result in selection of the benchmark with the lowest concentration. Additional information on the types of benchmarks and the benchmarks for all constituents analyzed is provided by Burton and Belitz (2008) and Shelton and others (2008).

Relative-concentrations were classified into high, moderate, and low categories:

Category	Relative-concentrations for organic constituents	Relative-concentrations for inorganic constituents
High	> 1	> 1
Moderate	> 0.1 and ≤ 1	> 0.5 and ≤ 1
Low	≤ 0.1	≤ 0.5

A relative-concentration greater than 1.0 is classified as high. A relative-concentration of 0.1 was used as a boundary between moderate and low values of organic and special-interest constituents for consistency with other studies and reporting requirements (Toccalino and others, 2004). The USEPA also established a relative-concentration of 0.1 of the regulatory benchmark as a threshold concentration so that the agency would be notified if the presence of a pesticide in surface water or groundwater is greater than or equal to that threshold (U.S. Environmental Protection Agency, 1997). In addition, organic and special-interest constituents, which generally are anthropogenic, usually are less prevalent and have smaller maximum relative-concentrations than inorganic constituents. In contrast, inorganic constituents are typically naturally occurring at concentrations that could be greater than 0.1 of regulatory benchmarks; consequently, it would be difficult to identify inorganic constituents that may have elevated concentrations greater than background levels if a relative-concentration of 0.1 was used as the threshold between moderate and low relative-concentrations. Therefore, the boundary between moderate and low relative-concentrations was set at 0.5 of the regulatory benchmark.

Design of Sampling Networks for Status Assessment

The wells selected for sampling by the USGS in the SESJ and KERN study units were selected to provide a statistically unbiased, spatially distributed set of wells for the assessment of the quality of groundwater in the primary aquifers (USGS-grid wells). Water-quality data from the USGS-grid wells were supplemented with data from selected wells from the CDPH database (CDPH-grid wells, and discussed in more detail in the "California Department of Public Health Grid Well Selection" section) to obtain more complete grid coverage and to include constituents that were not analyzed for in every USGS-grid well. These data were used to assess proportions of the primary aquifers with high, moderate, and low relative-concentrations.

The primary data used for the grid-based calculations of aquifer-scale proportions were data from wells sampled by the GAMA Priority Basin Project. Detailed descriptions of the methods used to identify wells for sampling are given in Burton and Belitz (2008) and Shelton and others (2008). USGS-grid wells (83 wells in the SESJ study unit and 47 wells in the KERN study unit) were selected to provide a statistically unbiased, spatially distributed set of wells for the assessment of the quality of groundwater in the primary aquifers (Scott, 1990). The objective of the grid design was to sample one CDPH well in each cell. If a grid cell did not contain accessible CDPH wells, then commercial, irrigation, or domestic wells were considered for sampling. The USGS-grid wells were sampled by the USGS for the GAMA Priority Basin Project, but are owned by other organizations or individuals.

One USGS-grid well was sampled in 83 of the 102 grid cells in the SESJ study unit, including 39 of the 40 grid cells in the Kings study area, 18 of the 20 grid cells in the Kaweah study area, 17 of the 20 grid cells in the Tule study area, and 9 of the 22 grid cells in the Tulare Lake study area (fig. 6). One USGS-grid well was sampled in 47 of the 122 grid cells in the KERN study unit (fig. 6). The grid cells from which samples were not collected had few, if any, wells, or permission to sample was not granted for wells that did exist in those cells. The 130 USGS-grid wells sampled in the southern San Joaquin Valley included 108 CDPH wells, 6 domestic wells, 15 irrigation wells, and 1 fire-protection well. The CDPH wells, irrigation wells, and the fire-protection well are considered production wells for this report. USGS-grid wells in the SESJ study unit were numbered in the order of sample collection with the prefix varying by study area: Kings study area (KING), Kaweah study area (KWH), Tule study area (TULE), Tulare Lake study area (TLR) (fig. A1A, appendix A). USGS-grid wells in the KERN study unit were numbered in the order of sample collection with the prefix KERN (fig. A2A, appendix A).

Samples collected from USGS-grid wells were analyzed for 180 to 345 constituents (table 1). VOCs, pesticides, perchlorate, noble gases, tritium, and stable isotopes of hydrogen and oxygen were analyzed in water samples from all wells. Additional pesticides, N-nitrosodimethylamine (NDMA), major and minor ions, trace elements, nutrients, isotopes of nitrogen and oxygen in nitrate, uranium isotopes, and redox species were analyzed in samples from 42 USGS-grid wells. Gasoline oxygenates, dissolved organic carbon, additional radiochemical constituents, carbon isotopes, and microbial constituents were analyzed in samples from 19 USGS-grid wells. The collection, analysis, and quality-control data for the analytes listed in table 1 are described by Burton and Belitz (2008) and Shelton and others (2008).

California Department of Public Health Grid Well Selection

Data for VOCs, pesticides, and perchlorate were collected at all 130 USGS-grid wells. The USGS-grid-well data included more VOC and pesticide constituents, and reporting levels were lower than reporting levels from the CDPH database. Therefore, CDPH data for these constituents were not used to supplement USGS-grid-well data for the status assessment.

Samples for analysis of inorganic constituents were collected from 42 of 130 USGS-grid wells (28 in the SESJ study unit, and 14 in the KERN study unit). Because the GAMA Priority Basin Project did not collect a complete suite of inorganic constituents for all grid cells, the CDPH database was used to provide data for inorganic constituents for the cells without this data (table 2). In addition, the GAMA Priority Basin Project was not able to sample wells in six of the grid cells. CDPH wells were selected to represent as many of these grid cells as possible. CDPH wells that were selected to supplement USGS-grid wells are referred to as "CDPH-grid" wells. The approach used to identify suitable CDPH wells is described in appendix A. Briefly, the first choice was to use CDPH data from the same well as the USGS-grid well ("DG" CDPH-grid wells; tables A1 and A2). If the DG well did not have all needed data, a second well was randomly selected from the subset of CDPH wells in the same cell with data ("DPH" CDPH-grid wells; tables A1 and A2). Combining data from CDPH-grid wells with data from USGS-grid wells produced inorganic data for 146 cells. All other CDPH wells with data from the current period (January 1, 2003, through December 31, 2005) not selected to be CDPH-grid wells are referred to as "CDPH-other" wells. Comparisons of data from USGS and CDPH wells to assess the validity of using these different sources in combination are presented in appendix B.

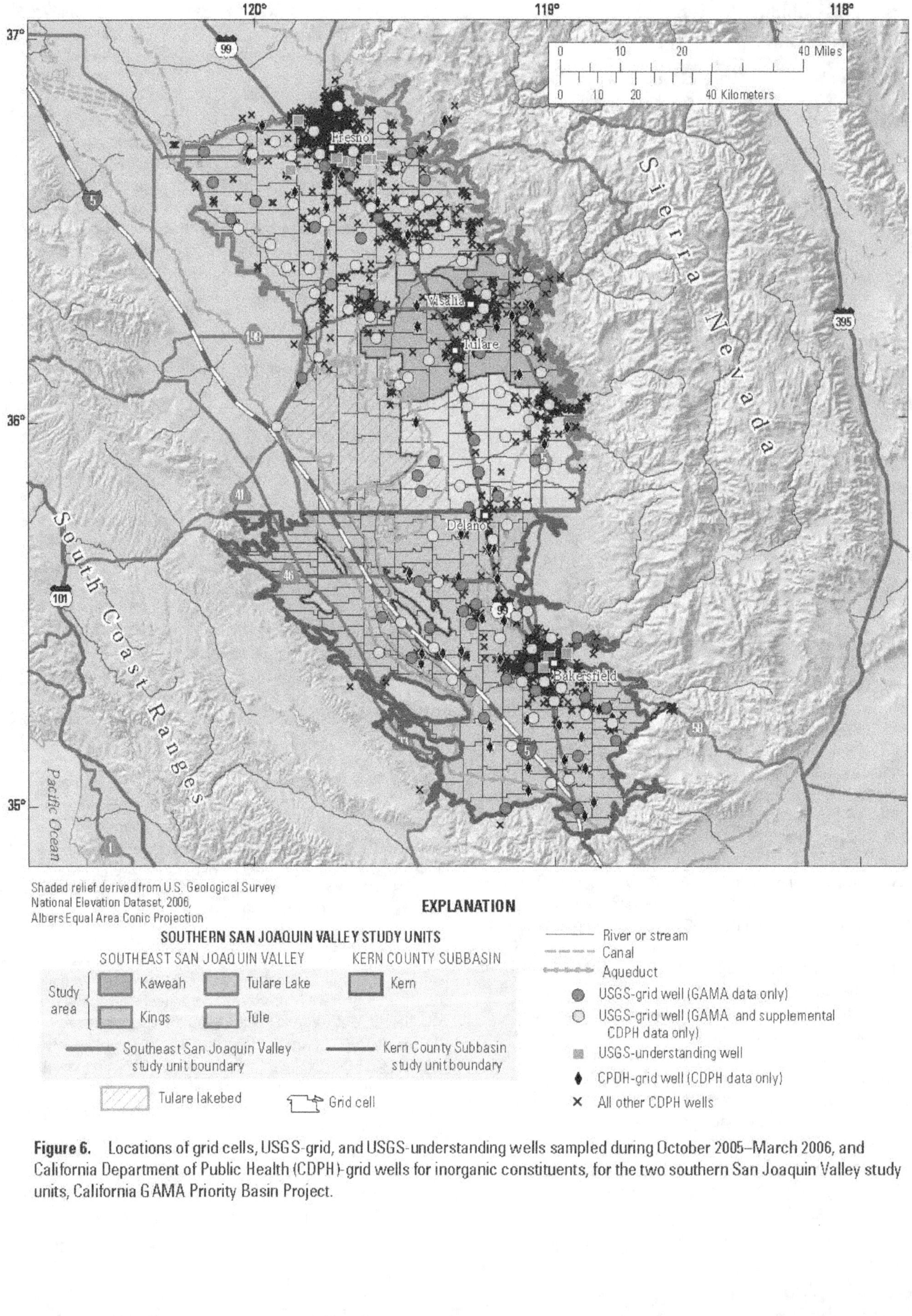

120°
119°
118°
37°
36°
35°
0 10 20 40 Miles
0 10 20 40 Kilometers
99
5
198
41
46
101
1
395
58
Fresno
Visalia
Tulare
Delano
Bakersfield
Sierra Nevada
South Coast Ranges
Pacific Ocean
Shaded relief derived from U.S. Geological Survey National Elevation Dataset, 2006, Albers Equal Area Conic Projection
EXPLANATION
SOUTHERN SAN JOAQUIN VALLEY STUDY UNITS
SOUTHEAST SAN JOAQUIN VALLEY
KERN COUNTY SUBBASIN
Study area
Kaweah
Tulare Lake
Kern
Kings
Tule
Southeast San Joaquin Valley study unit boundary
Kern County Subbasin study unit boundary
Tulare lakebed
Grid cell
River or stream
Canal
Aqueduct
USGS-grid well (GAMA data only)
USGS-grid well (GAMA and supplemental CDPH data only)
USGS-understanding well
CPDH-grid well (CDPH data only)
All other CDPH wells

Figure 6. Locations of grid cells, USGS-grid, and USGS-understanding wells sampled during October 2005–March 2006, and California Department of Public Health (CDPH)-grid wells for inorganic constituents, for the two southern San Joaquin Valley study units, California GAMA Priority Basin Project.

Table 1. Analytes and wells sampled for each analytical schedule for the two southern San Joaquin Valley study units, California GAMA Priority Basin Project, October 2005–March 2006.

[**Abbreviations**: GAMA, Groundwater Ambient Monitoring and Assessment Program; VOC, volatile organic compound; NDMA, *N*-nitrosodimethylamine; 1,2,3-TCP, 1,2,3-trichloropropane; μg/L, microgram per liter]

Well summary	Southeast San Joaquin Valley			Kern County Subbasin		
	Fast	Intermediate	Slow	Fast	Intermediate	Slow
	Number of wells					
Total number of wells	55	20	24	33	6	11
Number of grid wells sampled	55	19	9	33	4	10
Number of understanding wells sampled	0	1	15	0	2	1

Analyte group	Number of constituents	Schedule		
		Fast	Intermediate	Slow
Dissolved oxygen, specific conductance, temperature	3	X	X	X
Volatile organic compounds (VOCs) [1]	85	X	X	X
Pesticides and degradates	82	X	X	X
Perchlorate	1	X	X	X
Noble gases & tritium [2]	7	X	X	X
Stable isotopes of water	2	X	X	X
Pharmaceuticals [3]	12	X[11]	X	X
Laboratory alkalinity and pH	2		X	X
Polar pesticides and degradates [4]	54		X	X
NDMA and low-level 1,2,3-TCP [5]	2		X	X
Nutrients	5		X	X
Nitrogen and oxygen isotopes of nitrate	2		X	X
Major, minor, and trace elements [6]	36		X	X
Arsenic and iron speciation	4		X	X
Uranium isotopes	3		X	X
Tritium [7]	1		X	X
Gasoline oxygenates [8]	3			X
Dissolved organic carbon	1			X
Field alkalinity and pH	2			X
Carbon isotopes	2			X
Radon-222	1			X
Radium isotopes	2			X
Gross alpha and beta radioactivity[9]	4			X
Microbial constituents	4			X
Low-level halogenated VOCs (chlorofluorocarbons) [3,10]	25			X
Total number of constituents analyzed		180	301	345

[1] Includes 10 constituents classified as fumigants or fumigant synthesis byproducts.

[2] Analyzed at Lawrence Livermore National Laboratory, Livermore, California.

[3] Not discussed in this report.

[4] Does not include four constituents in common with pesticides and degradates.

[5] Includes one analyte, 1,2,3-TCP, in common with VOC analyses. However, the laboratory reporting level for the low-level analysis is 0.005 μg/L compared to 0.18 μg/L for the VOC analysis. Therefore, the low-level analysis is counted as a separate analysis.

[6] Includes one constituent, uranium, classified as a radioactive constituent later in this report.

[7] Analyzed at U.S Geological Survey Stable Isotope and Tritium Laboratory, Menlo Park, California.

[8] Does not include five constituents in common with VOCs.

[9] Both gross alpha and gross beta particle activities were measured after 72-hour and 30-day holding times; the 72-hour results are used in this report.

[10] Includes 22 analytes in common with VOC analyses. However, the laboratory reporting levels for the low-level analyses are two or three orders of magnitude lower than for the VOC analyses. Therefore, the low-level analyses are counted as separate analyses.

[11] Collected on the fast schedule only in the Kern County Subbasin study unit.

Table 2. Inorganic constituents, associated benchmark information, and number of grid wells per constituent for the two southern San Joaquin Valley study units, California GAMA Priority Basin Project.

[**Abbreviations**: GAMA, Groundwater Ambient Monitoring and Assessment Program; CDPH, California Department of Public Health; HAL-US, USEPA lifetime health advisory level; MCL-US, USEPA maximum contaminant level; MCL-CA, CDPH maximum contaminant level; NL-CA, CDPH notification level; AL-US, USEPA action level; SMCL-CA, CDPH secondary maximum contaminant level; SMCL-US, USEPA secondary maximum contaminant level; USEPA, U.S. Environmental Protection Agency]

Constituent	Benchmark type	Study unit			
		Southeast San Joaquin Valley		Kern County Subbasin	
		Number of grid wells sampled by GAMA	Number of grid wells selected from CDPH	Number of grid wells sampled by GAMA	Number of grid wells selected from CDPH
Nutrients with health-based benchmarks					
Ammonia, as nitrogen	HAL-US	28	0	14	0
Nitrite, as nitrogen	MCL-US	28	38	14	46
Nitrate, as nitrogen	MCL-US	28	51	14	53
Trace elements and minor ions with health-based benchmarks					
Aluminum	MCL-CA	28	44	14	42
Antimony	MCL-US	28	40	14	42
Arsenic	MCL-US	28	45	14	42
Barium	MCL-CA	28	44	14	42
Beryllium	MCL-US	28	44	14	42
Boron	NL-CA	28	18	14	18
Cadmium	MCL-US	28	44	14	42
Chromium	MCL-CA	28	43	14	40
Copper	AL-US	28	35	14	37
Lead	AL-US	28	41	14	39
Mercury	MCL-US	28	40	14	42
Molybdenum	HAL-US	28	0	14	0
Nickel	MCL-CA	28	44	14	42
Selenium	MCL-US	28	40	14	42
Strontium	HAL-US	28	0	14	0
Thallium	MCL-US	28	40	14	42
Vanadium	MCL-US	28	21	14	15
Fluoride	MCL-CA	28	39	14	43
Trace elements and major ions with secondary maximum contaminant levels					
Iron	SMCL-CA	28	33	14	39
Manganese	SMCL-CA	28	33	14	39
Silver	SMCL-CA	28	42	14	37
Zinc	SMCL-US	28	33	14	36
Chloride	SMCL-CA	28	33	14	34
Sulfate	SMCL-CA	28	33	14	34
Total dissolved solids	SMCL-US	51	33	24	34
Radioactive constituents with health-based benchmarks					
Gross alpha	MCL-US	9	44	10	28
Gross beta	MCL-US	9	1	10	3
Radon-222	MCL-US	9	0	10	1
Radium-226 + -228	MCL-US	9	24	9	9
Uranium	MCL-US	28	12	14	6

Selection of Constituents for Additional Evaluation

The GAMA Priority Basin Project used available monitoring data along with newly collected data for characterization of the groundwater resource. The statewide CDPH database contains data for regulated constituents with water-quality benchmarks. Although other organizations also collect water-quality data, the CDPH data is the only statewide database of public-supply well data available for comprehensive analysis. Data for some constituents, including VOCs, pesticides, inorganic constituents, and radioactive constituents, are available from the GAMA Priority Basin Project and the CDPH databases. However, more VOCs and pesticides are analyzed by the GAMA Priority Basin Project than were available in the CDPH database (table 3). In addition, laboratory reporting levels (LRLs) for GAMA Priority Basin Project data typically were one or two orders of magnitude less than the method detection levels (MDLs) used for analyses compiled by CDPH (table 3). Thus, the GAMA Priority Basin Project data was selected to enhance the CDPH database by providing a larger number of analytes and lower reporting levels than are found in the CDPH database. Both datasets are used in the status and understanding assessments.

The CDPH database contains more than 1,800,000 records from more than 2,200 wells in the SESJ and KERN study units, necessitating targeted retrievals to access water-quality data effectively. CDPH data were used with USGS-grid data to identify constituents in the study units at concentrations greater than water-quality benchmarks at any time during the period of record (March 28, 1980 through December 31, 2005, for the SESJ study unit and April 26, 1978 through December 31, 2005, for the KERN study unit). These constituents were included in the status assessment. Constituent concentrations retrieved from the CDPH database for samples in the study units were identified as "historically high" (table 4) if (1) concentrations were high (greater than benchmarks) at any time during the period of record and (2) concentrations were not high in the most recent 3-year period (January 1, 2003 through December 31, 2005, hereinafter referred to as current period) or in USGS-grid data. These constituents do not reflect current conditions on which the status assessment is based.

Table 3. Comparison of the number of compounds and median method detection limits or median laboratory reporting levels by type of constituent for data stored in the California Department of Public Health (CDPH) database and data collected by the U.S. Geological Survey in the two southern San Joaquin Valley study units, California GAMA Priority Basin Project, October 2005– March 2006.

[Median MDL and LRL units: VOCs, pesticides, pharmaceutical constituents, NDMA, perchlorate, and trace elements, micrograms per liter (µg/L); radioactive constituents, picocuries per liter (pCi/L); nutrients and major and minor ions, milligrams per liter (mg/L). **Abbreviations**: CDPH, California Department of Public Health; GAMA, Groundwater Ambient Monitoring and Assessment Program; MDL, method detection limit; LRL, laboratory reporting level; VOC, volatile organic compound; ssL_c, sample-specific critical level; nc, not collected]

Constituent	CDPH		GAMA	
	Number of compounds	Median MDL	Number of compounds	Median LRL
Organic constituents				
Volatile organic compounds (VOCs) plus gasoline oxygenates (including fumigants)	68	0.5	88	0.06
Pesticides plus degradates	68	1	136	0.016
Inorganic constituents				
Pharmaceutical constituents	nc	nc	14	[1] 0.024
Trace elements	19	10	25	0.12
Radioactive constituents (ssL_c)	6	2	11	[2] 0.66
Major and minor ions	10	unknown	11	0.10
Nutrients, dissolved organic carbon	6	0.4	6	0.05
Constituents of special interest				
Perchlorate	1	4	1	0.5
N-Nitrosodimethylamine (NDMA)	nc	nc	1	0.002

[1] Value reported is a median MDL.

[2] Value reported is a median sample-specific critical level (ssL_c) for 11 radioactive constituents collected and analyzed by GAMA.

Table 4.   Constituents in CDPH wells with historically high concentrations but not during the current period (January 1, 2003, to December 31, 2005) in the two southern San Joaquin Valley study units, California GAMA Priority Basin Project.

[Benchmark value units: trace elements, pesticides, fumigants, solvents, other organics, and constituents of interest, micrograms per liter (µg/L); radioactive constituents, picocuries per liter (pCi/L). A high analysis is defined as a concentration that is greater than the human-health benchmark for that constituent. **Abbreviations:** CDPH, California Department of Public Health; GAMA, Groundwater Ambient Monitoring and Assessment Program; MCL-US, USEPA maximum contaminant level; MCL-CA, CDPH maximum contaminant level; HAL-US, USEPA health advisory level; RSD5-US, USEPA risk-specific dose at a risk factor of 10^{-5}; USEPA, U.S. Environmental Protection Agency]

Constituent	Number of wells with analyses	Benchmark type	Benchmark value	Date of most recent high value	Number of wells with at least one historically high value
Southeast San Joaquin Valley study unit (March 28, 1980, to December 31, 2002)					
Trace elements					
Antimony	950	MCL-US	6	01-24-1994	1
Beryllium	962	MCL-US	4	06-01-2000	1
Cadmium	1,075	MCL-US	5	07-02-2002	9
Chromium	1,068	MCL-CA	50	08-28-2002	2
Mercury	1,062	MCL-US	2	05-12-1989	4
Molybdenum	6	HAL-US	40	07-24-2001	1
Radioactive constituents					
Radium-226	163	MCL-US	5	03-16-1995	2
Radium-228	275	MCL-US	5	07-01-1994	1
Pesticides					
Diazinon	844	HAL-US	1	06-17-1996	1
Dieldrin	514	RSD5-US	0.02	02-03-1982	1
Fumigants					
1,2-Dichloropropane	1,095	MCL-US	5	01-05-1994	2
1,4-Dichlorobenzene	1,144	MCL-CA	5	01-05-1994	1
Solvents					
1,1-Dichloroethane	1,144	MCL-CA	5	01-05-1994	1
1,2-Dichloroethane	1,144	MCL-CA	0.5	01-05-1994	1
trans-1,2-Dichloroethene	1,142	MCL-CA	10	07-05-1988	1
1,1,2,2-Tetrachloroethane	1,143	MCL-CA	1	01-05-1994	2
1,2,4-Trichlorobenzene	1,001	MCL-CA	5	01-05-1994	1
1,1,2-Trichloroethane	1,143	MCL-US	5	01-05-1994	1
Other organics					
Vinyl chloride	1,143	MCL-CA	0.5	01-05-1994	1
Kern County Subbasin study unit (April 26, 1978, to December 31, 2002)					
Trace Elements					
Aluminum	488	MCL-CA	1,000	05-01-2000	4
Barium	498	MCL-CA	1,000	12-19-2001	3
Chromium	498	MCL-CA	50	03-08-1993	2
Mercury	499	MCL-US	2	10-20-1999	5
Radioactive constituents					
Radium-226	119	MCL-US	5	11-05-1993	1
Gross beta radioactivity	56	MCL-CA	50	07-10-2002	2
Pesticides					
Atrazine	472	MCL-CA	1	08-04-1998	3
Solvents					
1,2-Dichloroethane	503	MCL-CA	0.5	07-16-2002	12
Constituent of special interest					
Perchlorate	251	MCL-CA	6	05-13-2001	1

More than 340 constituents were analyzed in the southern San Joaquin Valley study units; however, only a subset of these constituents is selected for additional evaluation in this report. Three criteria were used to identify constituents for additional evaluation:

1. Constituents with concentrations at high or at moderate relative-concentrations in the CDPH database during the current 3-year period (January 1, 2003, to December 31, 2005),

2. Constituents with concentrations at high or at moderate relative-concentrations in the USGS-grid wells or USGS-understanding wells, or

3. Organic constituents that were detected in more than 10 percent in the USGS-grid-well dataset, even if relative-concentrations were low.

The relative-concentrations for constituents discussed in the *understanding assessment* were high in more than 2 percent of the primary aquifers, or constituents were detected in more than 10 percent of the USGS-grid well dataset. A complete list of the constituents investigated by GAMA Priority Basin Project in the southern San Joaquin Valley study units may be found in the data reports for the study unit (Burton and Belitz, 2008; Shelton and others, 2008).

Calculation of Aquifer-Scale Proportions

The *status assessment* is intended to characterize the quality of groundwater resources in the primary aquifers of the two southern San Joaquin Valley study units. The primary aquifers are defined by the depth intervals over which wells listed in the CDPH database are perforated. The use of the term "primary aquifers" does not imply that there is a discrete aquifer unit. In most groundwater basins, municipal and community supply wells generally are perforated at greater depths than are domestic wells. Most of the wells used in the *status assessment* are listed in the CDPH databases. Thus, because domestic wells are not listed in the CDPH database, the primary aquifers generally correspond to the part of the aquifer system tapped by municipal and community supply wells.

Water quality in the primary aquifers can differ from water quality in shallow or deep parts of the aquifer system. Previous investigations in the study unit have shown that groundwater in shallow parts of the aquifer generally is of poorer quality than groundwater at greater depths in the aquifer (Burow and others, 1998a, 1998b, 2007, 2008b). Similarly, water quality at greater depths than those typically used for public supply can be of different quality, particularly with respect to dissolved solids (Page, 1973). The proportions for the primary aquifers discussed in this report do not characterize the shallow or deep parts of the aquifer system.

Two statistical methods—grid-based and spatially weighted—were applied to evaluate the proportions of the primary aquifers in the southern San Joaquin Valley study

units with high, moderate, and low relative-concentrations of constituents. For ease of discussion, these proportions are referred to as "high," "moderate," and "low" aquifer-scale proportions. Calculations of aquifer-scale proportions were made for individual constituents meeting the criteria for additional evaluation in the *status assessment*, and for classes of constituents. Classes of constituents with health-based benchmarks included trihalomethanes (THMs), solvents, other VOCs, fumigants, pesticides, trace elements and minor ions, uranium and radioactive constituents, and nutrients. Among constituents with aesthetic benchmarks (SMCLs), aquifer-scale proportions were calculated for major ions (total dissolved solids, chloride, and sulfate) in addition to manganese and iron.

The grid-based calculation uses the grid-well dataset assembled from the USGS- and CDPH-grid wells (Belitz and others, 2010). The proportion of the primary aquifers with high relative-concentrations of a constituent was calculated by dividing the number of grid cells represented by a high value for that constituent by the total number of grid cells with data for that constituent (see appendix C for details of methods). Proportions of moderate and low relative-concentrations were calculated similarly. Confidence intervals for grid-based detection frequencies of high relative-concentrations were computed using the Jeffreys interval for the binomial distribution (Brown and others, 2001). Although the grid-based estimate is spatially unbiased, the grid-based approach may not detect constituents that are present at high concentrations in small proportions of the primary aquifers. For calculation of high aquifer-scale proportion for a class of constituents, cells were considered high if the value for any of the constituents was high. Cells were considered moderate if the value for any of the constituents was moderate, but none of the values were high.

The spatially weighted calculation used all available data from the following sources to calculate the aquifer-scale proportions—(1) all CDPH wells in each study unit (most recent analysis from each well with data for the constituent during the current period, January 1, 2003, to December 31, 2005), (2) USGS-grid wells, and (3) USGS-understanding wells with perforation intervals representative of the primary aquifers (discussed in the section Understanding Assessment Methods). USGS-understanding wells that were monitoring wells were excluded because these wells were perforated at shallower depths than is typical for wells in the CDPH database. For the spatially weighted approach, proportions are computed on a cell-by-cell basis (Isaaks and Srivistava, 1989; Belitz and others, 2010), rather than as an average of all wells. The proportion of high relative-concentrations for each constituent for the primary aquifers was computed (1) by computing the proportion of wells with high relative-concentrations in each grid cell and (2) by averaging the grid-cell proportions computed in step (1) (see appendix C for details of methods). Similar procedures were used to calculate the aquifer-scale proportions of moderate and low relative-concentrations. The resulting proportions are spatially unbiased (Isaaks and Srivastava, 1989).

Detection frequencies of wells with high relative-concentrations for constituents calculated using the same data that was used for the spatially weighted approach are provided for reference in this report, but were not used to assess aquifer-scale proportions. Detection frequencies are not spatially unbiased because the wells in the CDPH database are not uniformly distributed. Consequently, high relative-concentrations in spatially clustered wells in a particular area representing a small part of the primary aquifers could be given a disproportionately high weight compared to spatially unbiased methods.

The grid-based aquifer-scale proportions were used to represent proportions in the primary aquifers unless the spatially weighted proportions were significantly different than the grid-based values. Significantly different results were defined as follows:

1. If the grid-based high aquifer-scale proportion was zero and spatially weighted aquifer-scale proportion was non-zero, then the spatially weighted result was used. This situation can arise when the concentration of a constituent is high in a small fraction of the aquifer.

2. If the grid-based high aquifer-scale proportion was non-zero, then the 90 percent confidence interval (based on the Jeffreys interval for the binomial distribution, Brown and others, 2001) was used to evaluate the difference. If the spatially weighted proportion was outside the 90 percent confidence interval, then the spatially weighted proportion was used.

The grid-based moderate and low proportions were used in most cases because the reporting limits for many organic constituents and some inorganic constituents in the CDPH database were higher than the boundary between the moderate and low categories. However, if the grid-based moderate proportion was zero and the spatially weighted proportion non-zero, then the spatially weighted value was used as an estimate for the moderate proportion.

Understanding Assessment Methods

The potential explanatory factors—land use, well depth, depth to the top-of-perforations, normalized position of wells along flow paths, classified groundwater age, and geochemical condition (see appendix D for more details)—were analyzed in relation to constituents selected for additional evaluation for the understanding assessment in order to establish context for physical and chemical processes within the groundwater system. Statistical tests were used to identify significant correlations between the constituents of interest and potential explanatory factors. Graphs, bar charts, and maps were used to improve the understanding of factors affecting water quality for selected correlations.

The wells selected for the understanding assessment were USGS- and CDPH-grid wells and USGS-understanding wells. CDPH-other wells were not used in the understanding assessment because carbon isotope, tritium, dissolved oxygen, and some well construction data were not available.

U.S. Geological Survey Understanding Wells

Nineteen wells (16 wells in the SESJ study unit and 3 wells in the KERN study unit) were sampled to improve understanding of factors and processes that affect groundwater quality. The USGS-understanding wells sampled in the study units were numbered in the order of collection with a prefix modified from those used for the USGS-grid wells. USGS-understanding wells sampled in the SESJ study unit were designated as either *flow-path* (for example, KING-*FP*) or *transect wells* (HWY99*T*). All three USGS-understanding wells in the KERN study unit were designated as *flow-path wells*. The understanding wells included 11 monitoring and 8 CDPH wells. The USGS-understanding wells were selected (1) to assess changes in water quality along regional groundwater flow paths from east to west across the Kings and Kern study areas; (2) to compare water quality at depths less than 250 ft where most of the USGS-understanding wells are perforated with water quality at depths greater than 250 ft where CDPH wells generally are perforated [previous investigations have identified that vertical changes in water chemistry occur, primarily within relatively shallow to intermediate depths, with more uniform water quality at greater depths in the aquifer system (Burow and others, 2007, 2008b; Jurgens and others, 2008)]; or (3) to assess differences in water quality along an approximate regional transect across various alluvial fans in the southern San Joaquin Valley, roughly paralleling Highway 99.

Statistical Analysis

Nonparametric statistical methods were used to test the significance of correlations between water-quality parameters and potential explanatory variables. Nonparametric statistics are robust techniques that generally are not affected by outliers and do not require that the data follow any particular distribution (Helsel and Hirsch, 2002). The significance level (p) used for hypothesis testing for this report was compared to a threshold value (α) of 5 percent ($\alpha=0.05$) to evaluate whether the relation was statistically significant ($p<\alpha$). Two different types of statistical tests were used because the set of potential explanatory factors included both continuous and categorical variables. Relations between categorical variables (for example, classified groundwater age or land-use class) and water-quality variables were evaluated using the Wilcoxon rank-sum or Kruskal-Wallis nonparametric tests. Correlations between continuous variables were evaluated using Spearman's method. Correlations between potential explanatory factors, between water-quality parameters, and between potential explanatory factors and water-quality constituents were tested for significance. Correlations of total THMs were performed on the sum of THM concentrations. For example, the total THM concentrations for KWH-17 is 1.36 micrograms per liter (µg/L) [0.42 (chloroform) + 0.39 (bromodichloromethane) + 0.15 (bromoform) + 0.4 (dibromochloromethane)].

Correlations between explanatory factors and groundwater constituents were tested using either the set of USGS- and CDPH-grid plus understanding wells or USGS- and CDPH-grid wells only. Because the USGS-understanding wells primarily represented relatively shallow groundwater in agricultural areas that were not randomly selected on a spatially distributed grid, they were excluded from analysis of relations between water quality and areally distributed explanatory variables (land use and lateral position) to avoid areal-clustering bias. However, USGS-understanding wells were included in analysis of relations between water-quality constituents and vertically distributed explanatory factors (depth, classified groundwater age, and oxidation-reduction characteristics). In addition, wells located in the SESJ study unit that were sampled as part of the Domestic Well Program conducted by SWRCB were included in the analysis where applicable.

Potential Explanatory Factors

A brief description of potential explanatory factors including land use, well depth, normalized lateral position in the flow system, septic-system density, formerly leaking underground fuel tanks (LUFTs), groundwater age, and geochemical conditions are described in this section. The data sources and methodology used for assigning values for potential explanatory factors are described in appendix D.

Land Use

In the southern San Joaquin Valley, land use is a combination of agricultural, urban, and natural; however, land use in the areas surrounding the southern San Joaquin Valley primarily is natural (fig. 4). Land use in the SESJ study unit is 85 percent agricultural, 6 percent urban, and 9 percent natural (fig. 3B). Within the 500-meter (m) (1,640-ft) radius around each USGS- and CDPH-grid well, average land use was 57 percent agricultural (lower than for the SESJ study unit) and urban and natural average land use were higher than for the study unit at 31 and 12 percent, respectively. In contrast, within the 500-m (1,640-ft) radius around each CDPH well (CDPH-grid and CDPH-other wells), the average land use was 40 percent agricultural (lower than for the SESJ study unit or the grid wells), and urban average land use was higher than for the study area or grid wells at 48 percent (fig. 3B). In general, the land use around the grid wells in the SESJ study

unit over-represent the urban land use and under-represent the agricultural land use but are closer to the overall land use of the study unit than the land use around the CDPH wells.

Land use in the KERN study unit is 66 percent agricultural, 3 percent urban, and 31 percent natural. Within the 500-m (1,640-ft) radius around each USGS- and CDPH-grid well, average land use was 58 percent agricultural (lower than for the KERN study unit), and natural and urban average land use was higher at 23 and 19 percent, respectively. In contrast, within the 500-m (1,640-ft) radius around each CDPH well (CDPH-grid and CDPH-other wells), the average land use was 37 percent agricultural (lower than for the KERN study unit or the grid wells), and urban average land use was higher than for the study unit or grid wells at 44 percent (fig. 3B). Similar to the SESJ study unit, the grid wells in the KERN study unit over-represent the urban land use and under-represent the agricultural land use but are closer to the overall land use of the study unit than the land use around the CDPH wells.

Average land use within 500-m (1,640-ft) radius around each domestic well that was sampled by SWRCB as part of the Domestic Well Project was 80 percent agricultural, 5 percent urban, and 15 percent natural. These percentages were very similar to the land-use percentages in the Kaweah and Tule study areas where most of the domestic wells sampled were located (fig. A3). Land use in the Kaweah study area was 82 percent agricultural, 5 percent urban, and 13 percent natural, whereas land use in the Tule study area was 88 percent agricultural, 3 percent urban, and 9 percent natural.

The percentage of agricultural land use was adequate for correlation with most water-quality constituents. However, in some cases, constituents needed to be correlated with the percentage of orchard or vineyard land use (a subset of agricultural land use) to improve understanding of the relation between water quality and explanatory factors. The percentage of orchard or vineyard land use within the 500-m radius around wells is presented in tables D2, D3, and D4.

Depth

The depth of USGS- and CDPH-grid wells varied between the two study units. The median well depth for SESJ grid wells was 515 ft (157 m) below land surface; well depths ranged from 76 to 1,641 ft (23–500 m; fig. 7A). The median well depth for KERN grid wells was 719 ft (219 m) below land surface; well depths ranged from 400 to 1,496 ft (122–456 m; fig. 7B). A similar pattern was observed for depth to the top-of-perforations. The median depth to top-of-perforations for SESJ grid wells was 245 ft (75 m), and the median depth to the top-of-perforations for KERN grid wells was 395 ft (120 m). These values represent a subset of the grid wells because well depth and depth to the top-of-perforations were not known for several wells. Only wells with construction information available for the study units (tables A1 and A2) were included in the analyses of explanatory variables involving depth.

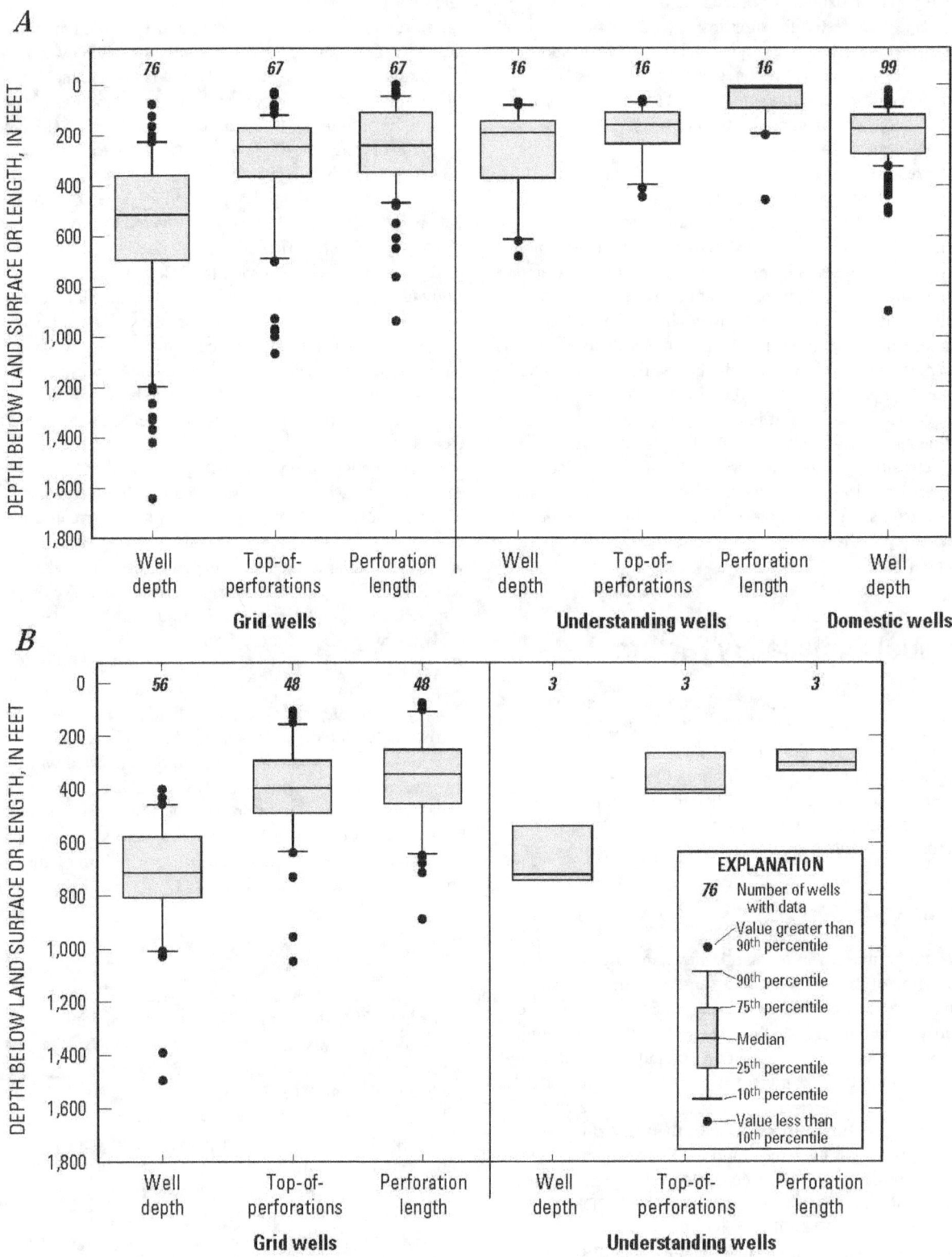

Figure 7.   Boxplots of construction characteristics for USGS-grid and USGS-understanding wells, and wells from the Domestic Well Project sampling in (*A*) the Southeast San Joaquin Valley study unit and (*B*) the Kern County Subbasin study unit, California GAMA Priority Basin Project.

The median depth of USGS-understanding wells (190 ft, 58 m) in the SESJ study unit was shallower than the median depth for grid wells (fig. 7A). This was expected because many of the understanding wells were monitoring wells that tap shallow groundwater. In contrast, the median depth of USGS-understanding wells (700 ft, 213 m) in the KERN study unit was similar to the median depth for grid wells (fig. 7B). This was expected because these understanding wells were additional CDPH wells.

The median depth of the 143 domestic wells (168 ft, 51 m) sampled for the Domestic Well Project located in the SESJ study unit was much shallower than the median depth of the SESJ grid wells (515 ft, 157 m) but is similar to the median depth for the understanding wells in the SESJ study unit (fig. 7A). The maximum depth of the domestic wells (900 ft, 274 m) was not as deep as the maximum depth of the SESJ grid wells (1,641 ft, 500 m) but was deeper than any of the understanding wells. The Domestic Well Project did not sample wells in the KERN study unit.

Lateral Position in the Flow System

USGS- and CDPH-grid wells were distributed across the entire range of normalized lateral positions (figs. 8 and 9A and B). Wells in the SESJ study unit with lateral positions of 0.0 to 0.20 (distal or western) and greater than 0.80 (proximal or eastern) made up 20 percent and 28 percent of the total grid wells, respectively (fig. 9A). USGS- and CDPH-grid wells in the KERN study unit with lateral positions of 0.0 to 0.20 (distal or western) and greater than 0.80 (proximal or eastern) made up 25 percent and 6 percent of the total grid wells, respectively (fig. 9B). The KERN study unit has fewer USGS- and CDPH-grid wells in the eastern part than the SESJ study unit. Lateral position for each grid well in the study units can be found in tables D2 and D3.

Domestic wells were not distributed across the entire range of lateral positions (figs. 8 and 9C). The majority of the domestic wells (57 percent) were located near the eastern boundary (lateral position greater than 0.80) of the SESJ study unit. Less than 1 percent had lateral position of 0.0 to 0.20.

Septic-System Density

The number of septic tanks or cesspools in the 500-m (1,640-ft) radius around each USGS-grid and understanding well in the SESJ study unit ranged from 0 to 71 septic tanks, with a median of 5 septic tanks. Septic-system density greater than the median value occurred in a larger fraction of grid wells in the Kaweah study area (15 of 26, 58 percent) than in the other study areas (34 of 81, 42 percent) (table D2). The number of septic tanks or cesspools in the 500-m (1,640-ft)

radius around each USGS-grid and understanding well in the KERN study unit ranged from 0 to 218 septic tanks, with a median of 1 septic tank (table D3). The number of septic tanks or cesspools in the 500-m radius around each domestic well sampled for the Domestic Well Project ranged from 0.2 to 101 septic tanks, with a median of 5.8 septic tanks (table D4).

Formerly Leaking Underground Fuel Tanks

The density of LUFTs located within the Thiessen polygon (a description of a Thiessen polygon can be found in appendix D in the section Formerly Leaking Underground Fuel Tanks) around each USGS- and CDPH-grid and USGS-understanding well in the SESJ study unit ranged from 0.01 to 9.23 tanks per square kilometer (tanks/km^2), with a median of 0.05 tanks/km^2 (table D2). The density of LUFTs around grid and understanding wells in the KERN study unit ranged from 0.004 to 4.17 tanks/km^2, with a median of 0.02 tanks/km^2 (table D3). The LUFT density for most of the wells in both study units was very low, usually less than 0.1 tank/km^2.

Groundwater Age

Groundwater samples were assigned age classifications based on the tritium, carbon-14, and helium-4 content of the samples (appendix D). Of the 99 USGS-grid and understanding wells in the SESJ study unit sampled by the Priority Basin Project, groundwater samples were classified as modern age in 38 wells, mixed age in 23 wells (evidence of both modern and pre-modern groundwater in the same sample), and pre-modern age in 36 wells (table D5). Samples from two wells could not be classified because the age-tracer data were incomplete.

The median depth of USGS-grid and -understanding wells in the SESJ study unit classified as pre-modern age was deeper than the depths of wells classified as modern or mixed ages (table 5). The median depth to the top-of-perforations of wells classified as pre-modern age also was deeper than the depth to the top-of-perforations of wells classified modern or mixed ages (fig. 10A, table 5). Well depths or depths to top-of-perforations in samples with modern and mixed ages were not significantly different.

Groundwater ages for nearly all SESJ wells perforated entirely at depths less than 250 ft (76 m) below land surface (19 of 20 wells) were modern or mixed (fig. 10B). Likewise, groundwater ages for most wells perforated entirely at depths greater than or equal to 250 ft (23 of 29 wells) were pre-modern. Groundwater ages for most of the wells with the top-of-perforation less than 250 ft but with the bottom-of-perforation greater than or equal to 250 ft were modern or mixed (18 of 22 wells).

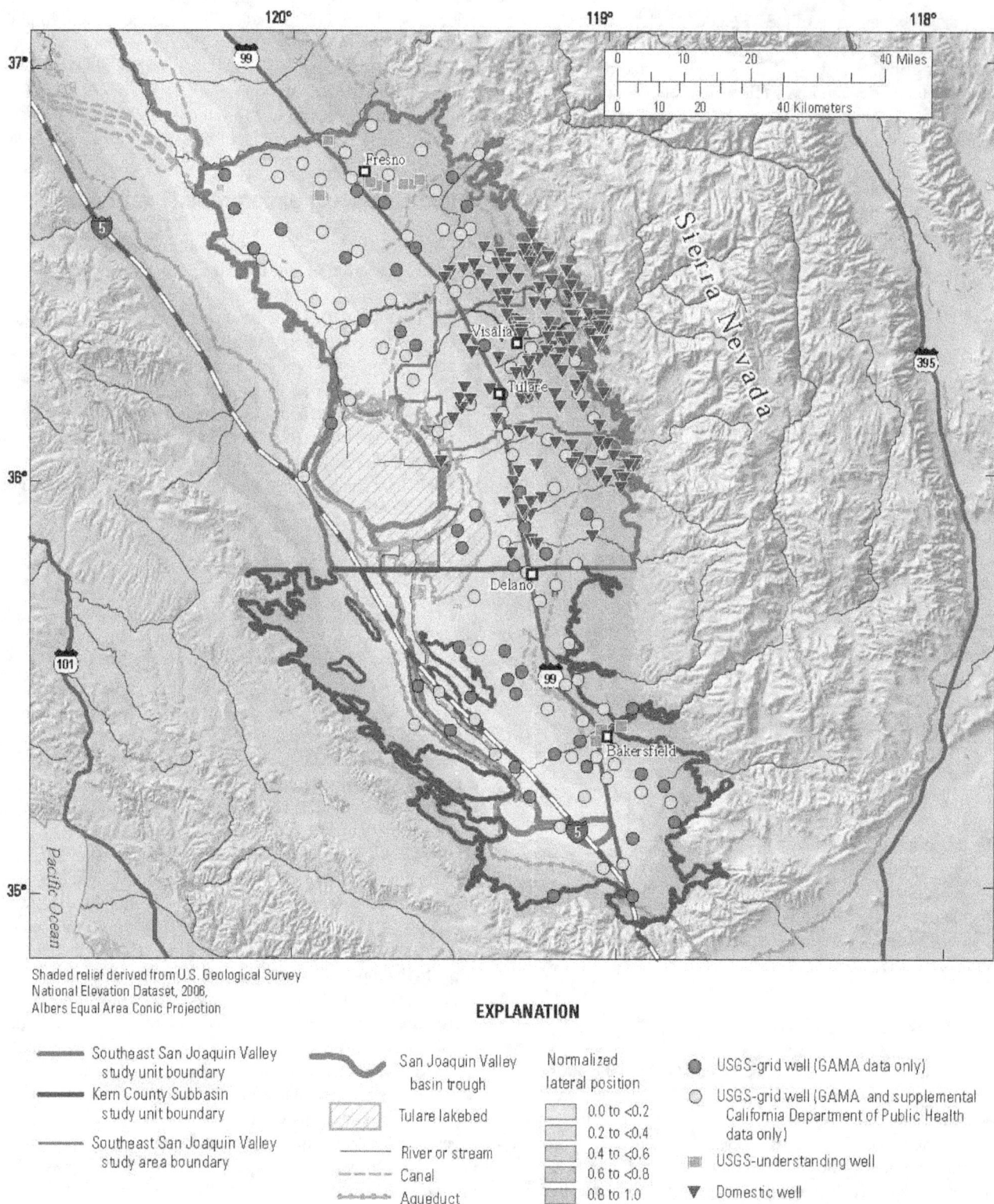

Shaded relief derived from U.S. Geological Survey
National Elevation Dataset, 2006,
Albers Equal Area Conic Projection

EXPLANATION

—— Southeast San Joaquin Valley
      study unit boundary

—— Kern County Subbasin
      study unit boundary

—— Southeast San Joaquin Valley
      study area boundary

〰️ San Joaquin Valley
      basin trough

▨ Tulare lakebed

—— River or stream

– – – Canal

~~~ Aqueduct

Normalized
lateral position

▢ 0.0 to <0.2
▢ 0.2 to <0.4
▢ 0.4 to <0.6
▢ 0.6 to <0.8
▢ 0.8 to 1.0

● USGS-grid well (GAMA data only)

○ USGS-grid well (GAMA and supplemental
   California Department of Public Health
   data only)

▦ USGS-understanding well

▼ Domestic well

Figure 8.   Normalized lateral position (distance from valley trough to valley margins) and wells, Southern San Joaquin Valley study units, California GAMA Priority Basin Project.

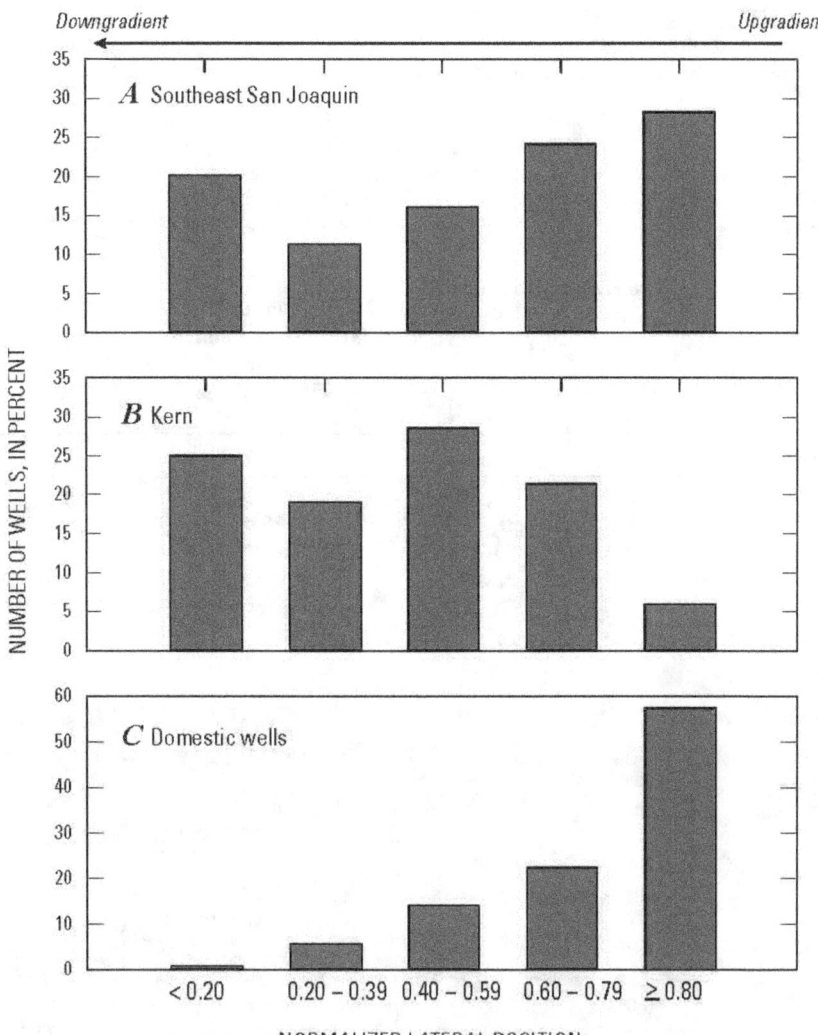

Figure 9.   Bar charts of distribution of USGS- and CDPH-grid wells and domestic wells for normalized lateral position, (A) Southeast San Joaquin Valley study unit, (B) Kern County Subbasin study unit, and (C) Domestic Well Project wells, California GAMA Priority Basin Project.

Table 5. Results of statistical tests (Kruskal-Wallis or Wilcoxon) for differences between selected potential categorical and continuous explanatory factors and differences between categorical explanatory factors and selected water-quality constituents for the two southern San Joaquin Valley study units, California GAMA Priority Basin Project, and the GAMA Domestic Well Project.

[Groundwater age class: Mod, modern; Mix, mixture of modern and pre-modern; preM, pre-modern (see appendix D for explanation of groundwater age classes). p-values less than 0.05 calculated using the Kruskal-Wallis (for comparing three or more groups) or Wilcoxon (for comparing two groups). Non-parametric tests indicate significant differences. **Abbreviations:** SESJ, Southeast San Joaquin Valley study unit; KERN, Kern County Subbasin study unit; DomWell, Domestic Well Project; Kings, Kings study area; KWH, Kaweah study area; TLR, Tulare Lake study area; Tule, Tule study area; Kern, Kern study area; VOC, volatile organic compounds; DBCP, 1,2-dibromo-3-chloropropane; ns, not significant; <, less than; >, greater than]

Potential explanatory factors	Groundwater age class[1] (Mod, Mix, preM) p-value significant differences[2]		Study unit (SESJ, KERN, DomWell) p-value significant differences[2]	Study area (Kings, KWH, TLR, TULE, Kern) p-value significant differences[2]
	SESJ	KERN		
Potential explanatory factors				
Well depth	<0.001 preM>Mod,Mix	0.033 preM>Mod	<0.001 KERN>SESJ>DomWell	<0.001 Kern,Tule,TLR>Kings,KWH
Depth to top-of-perforations[1]	<0.001 preM>Mod,Mix	0.016 preM>Mod	<0.001 KERN>SESJ	<0.001 Kern,TLR>Kings,KWH
Orchard or vineyard land use (percent)	ns	0.006 preM>Mod	<0.001 DomWell>SESJ,KERN	0.022 KWH>TLR,Kern
Agricultural land use (percent)[3]	ns	0.032 preM>Mod	<0.001 DomWell>SESJ,KERN	ns
Urban land use (percent)[3]	ns	ns	<0.001 SESJ>KERN>DomWell	0.029 Kings>Kern
Natural land use (percent)[3]	ns	ns	ns	ns
Normalized lateral position from valley trough	ns	ns	<0.001 DomWell>SESJ>KERN	<0.001 Kings, KWH>TLR,Kern; Tule>TLR
Number of septic tanks or cesspools	ns	ns	<0.001 SESJ,DomWell>KERN	<0.001 Kings,KWH>Kern
Number of formerly leaking underground fuel tanks[1]	ns	ns	ns	0.024 Kings,KWH>Tule,Kern
Dissolved oxygen (mg/L)[1]	<0.001 Mod>preM	0.009 Mod>preM	0.039 SESJ>KERN	<0.001 KWH>TLR,Kern
pH	0.041 preM>Mod	ns	<0.001 SESJ,KERN>DomWell	<0.001 TLR>Kings
Selected inorganic water-quality constituents				
Arsenic	0.007 preM>Mod	ns	<0.001 SESJ,KERN>DomWell	<0.001 TLR>Kings,KWH,Kern
Antimony	ns	ns	<0.001 KERN>DomWell	0.004 Kern>Kings
Boron	0.039 preM>Mix	ns	<0.001 SESJ,KERN>DomWell	0.003 TLR,Kern>Kings
Vanadium	ns	ns	<0.001 SESJ,DomWell>KERN	<0.001 Kings>TLR,Kern
Fluoride	0.003 preM>Mod	0.002 preM>Mod	<0.001 SESJ,KERN>DomWell	<0.001 TLR>Kings
Uranium	0.001 Mod,Mix>preM	ns	ns	ns
Nitrate	ns	ns	<0.001 DomWell>SESJ,KERN	<0.001 KWH>Tule>TLR,Kern
Manganese	0.006 preM>Mix,Mod	ns	<0.001 DomWell>SESJ,KERN	ns

Table 5. Results of statistical tests (Kruskal-Wallis or Wilcoxon) for differences between selected potential categorical and continuous explanatory factors and differences between categorical explanatory factors and selected water-quality constituents for the two southern San Joaquin Valley study units, California GAMA Priority Basin Project, and the GAMA Domestic Well Project.—Continued

[Groundwater age class: Mod, modern; Mix, mixture of modern and pre-modern; preM, pre-modern (see appendix D for explanation of groundwater age classes). p-values less than 0.05 calculated using the Kruskal-Wallis (for comparing three or more groups) or Wilcoxon (for comparing two groups). Non-parametric tests indicate significant differences. **Abbreviations**: SESJ, Southeast San Joaquin Valley study unit; KERN, Kern County Subbasin study unit; DomWell, Domestic Well Project; Kings, Kings study area; KWH, Kaweah study area; TLR, Tulare Lake study area; Tule, Tule study area; Kern, Kern study area; VOC, volatile organic compounds; DBCP, 1,2-dibromo-3-chloropropane; ns, not significant; <, less than; >, greater than]

Potential explanatory factors	Groundwater age class[1] (Mod, Mix, preM) p-value significant differences[2]		Study unit (SESJ, KERN, DomWell) p-value significant differences[2]	Study area (Kings, KWH, TLR, TULE, Kern) p-value significant differences[2]
	SESJ	KERN		
Selected inorganic water-quality constituents—Continued				
Iron	0.048 PreM>Mod	0.022 PreM>Mod	ns	0.049 TLR>KWH,Kings
Total dissolved solids (TDS)	ns	ns	<0.001 KERN,DomWell>SESJ	0.013 Kern>Kings,KWH,Tule
Sulfate	0.024 Mix>PreM	ns	<0.001 KERN>DomWell>SESJ	<0.001 Kern>Kings,KWH
Chloride	ns	ns	<0.001 KERN>SESJ,DomWell	<0.001 Kern,TLR>Kings,KWH,Tule
Selected organic and special-interest water-quality constituents				
Number of VOC detections [4]	<0.001 Mod>PreM	ns	<0.001 SESJ,KERN>DomWell	<0.001 KWH>Kings,Kern
Total trihalomethane (THM) concentration	0.006 Mod>PreM	ns	ns	<0.001 KWH>Kings,Tule
Number of solvent detections	0.028 Mod>PreM	ns	<0.001 SESJ>DomWell	ns
Number of other VOC detections	ns	0.024 PreM>Mod	<0.001 SESJ>DomWell	ns
DBCP	ns	ns	0.042 SESJ>KERN	ns
Number of pesticide detections [1]	<0.001 Mod,Mix>PreM	0.012 Mod>PreM	ns	0.024 KWH>TLR
Perchlorate	ns	ns	<0.001 DomWell>KERN	<0.001 KWH>TLR

[1] Data were not available for the domestic wells sampled as part of the Domestic Well Project.

[2] Only significant differences are shown. For example, PreM>Mod for well depth means that wells with pre-modern age water are significantly deeper than wells with modern age water, but wells with mixed age water are not significantly different from wells with either pre-modern or modern age water.

[3] Grid wells only.

[4] Does not include VOCs classified as fumigants.

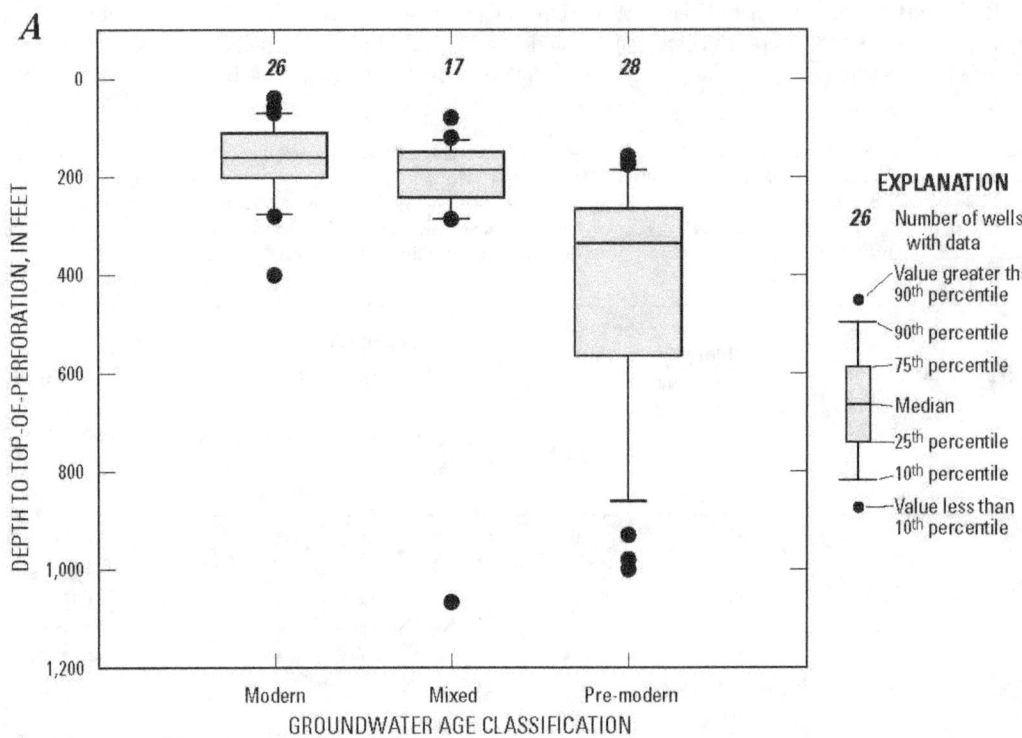

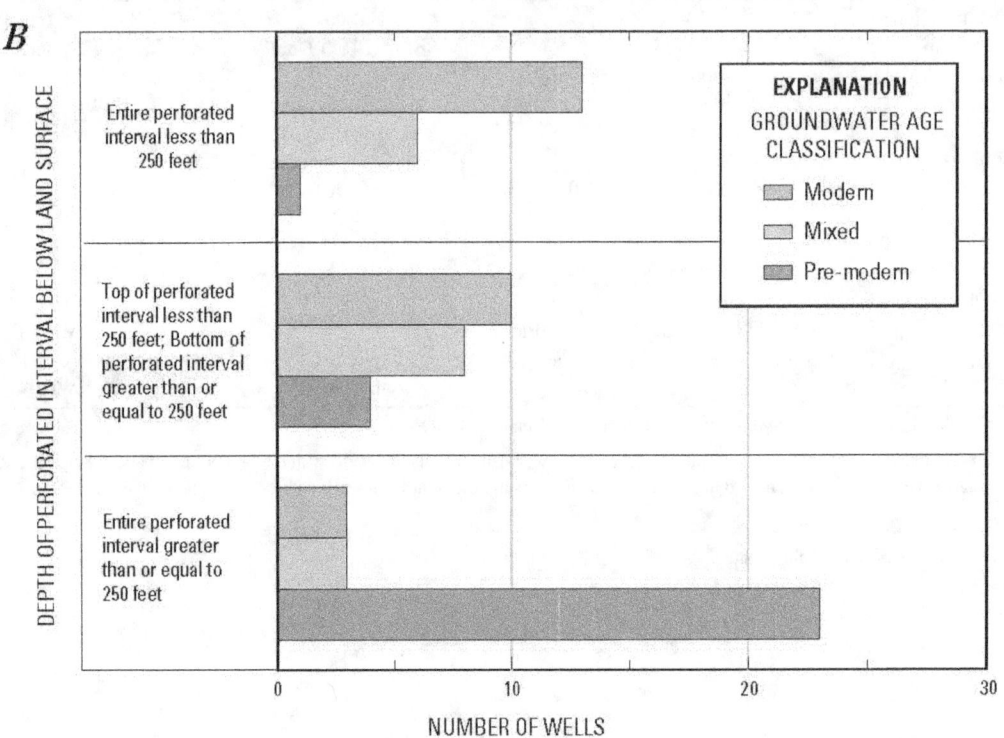

Figure 10. Boxplots and bar charts showing the relation of groundwater age classification to (*A*) depth to top-of-perforations and (*B*) numbers of wells with each groundwater age class in each of the three depth categories, Southeast San Joaquin Valley study unit, California GAMA Priority Basin Project.

In each of the three depth categories, ages for some wells are modern, mixed, and pre-modern. The presence of one pre-modern sample from one well less than 250 ft (76 m) deep and three modern samples from wells greater than or equal to 250 ft deep (fig. 10B) indicates that there are local variations in the general groundwater age-depth relations. These variations may indicate the position of the well relative to regional recharge and discharge areas.

Of the 50 USGS-grid and -understanding wells in the KERN study unit sampled by the Priority Basin Project, groundwater ages were classified as modern in 16 wells, mixed in 16 wells, and pre-modern in 14 wells (table D6). Samples from four wells could not be classified because the age-tracer data were incomplete or did not meet all quality-assurance checks.

Similar to the SESJ study unit, the median depth of USGS-grid and -understanding wells in the KERN study unit classified as pre-modern age was deeper than the median depths of wells classified as modern age (table 5). Depths of USGS-grid and -understanding wells classified as mixed age were not significantly different than depths of wells classified as modern or pre-modern (table 5). The median depth to the top-of-perforations of wells classified as pre-modern age also was deeper than the median depth to the top-of-perforations of wells classified as modern age (fig. 11A; table 5).

The water table below land surface in the KERN study unit is deeper than the water table in the SESJ study unit (Faunt, 2009). As a result, wells in the Kern study unit generally are deeper than in the SESJ study unit (fig. 8); therefore, a depth below land surface threshold of 500 ft (152 m) was used to categorize well depth and groundwater age in the Kern study unit rather than the depth of 250 ft (76 m) used in the SESJ study unit. Groundwater age for most of the wells perforated entirely between land surface and 500 ft below land surface (5 of 7 wells) was modern or mixed (fig. 11B). Likewise, nearly all of the wells perforated entirely at depths greater than or equal to 500 ft (6 of 7 wells) were of pre-modern age. The groundwater ages of most wells with the top-of-perforations less than 500 ft but with bottom-of-perforations greater than or equal to 500 ft were modern or mixed ages (17 of 22 wells).

Geochemical Conditions

Geochemical conditions investigated as potential explanatory factors in this report include oxidation-reduction characteristics and pH. An abridged classification of oxidation-reduction (redox) conditions adapted from the framework presented by McMahon and Chapelle (2008) for

USGS-grid and USGS-understanding wells in the southern San Joaquin Valley is given in appendix D (table D7 and D8). DO and pH were measured at USGS-grid and USGS-understanding wells. pH measurements for many of the CDPH-grid wells were available in the CDPH database, but no DO concentration data were available.

Groundwater in the SESJ study unit was oxic (redox category oxic or DO above 2.0 mg/L) in 73 percent of USGS-grid wells and 88 percent of USGS-understanding wells, but becomes more reducing (DO less than 2.0 mg/L) near the western (downgradient) area of the study unit (fig. 12). The lateral position and depth of wells having DO of 0.5 to 2.0 mg/L is consistent with general transitions from east to west and with increasing depth. Figure 12 shows the USGS-grid and -understanding wells with well-construction and DO data from the Kings study area on a single composite cross-section.

Groundwater in the KERN study unit was oxic (redox category oxic or DO above 2.0 mg/L) in 64 percent of the USGS-grid wells. In contrast to the SESJ study unit, measurements indicated reducing conditions (DO less than 2.0 mg/L) in groundwater in the KERN study unit in the eastern (upgradient) part of the study unit as well in the western (downgradient) part of the flow system (fig. 13). Reducing conditions (table D8) also were indicated in USGS-grid wells (KERN-32, -40, -43, and -46; fig. A2A) in the southern part of the KERN study unit, near the Tehachapi Mountains. High DO concentrations in the central part of the KERN study unit may indicate infiltration from the Kern River and groundwater banking facilities. Figure 13 shows the USGS-grid and USGS-understanding wells with well-construction and DO data which are adjacent to or north of the Kern River, on a single composite cross-section. Redox conditions for the CDPH-grid wells and the domestic wells from the Domestic Well Project were not categorized because DO data were not available for the wells.

Although the redox classification for the study units is valuable for characterizing the range and spatial distribution of redox conditions, hereinafter DO concentrations are used as the factor for evaluating relations of redox conditions with concentrations of water-quality constituents. DO was used as a redox indicator because (1) data were available for all USGS-grid and USGS-understanding wells, whereas other redox indicators were available for only 79 of the 130 USGS-grid wells, and (2) most groundwater samples in the southern San Joaquin Valley study units were classified as oxic, so the number of wells within the various reducing redox processes generally were too small for meaningful statistical analysis.

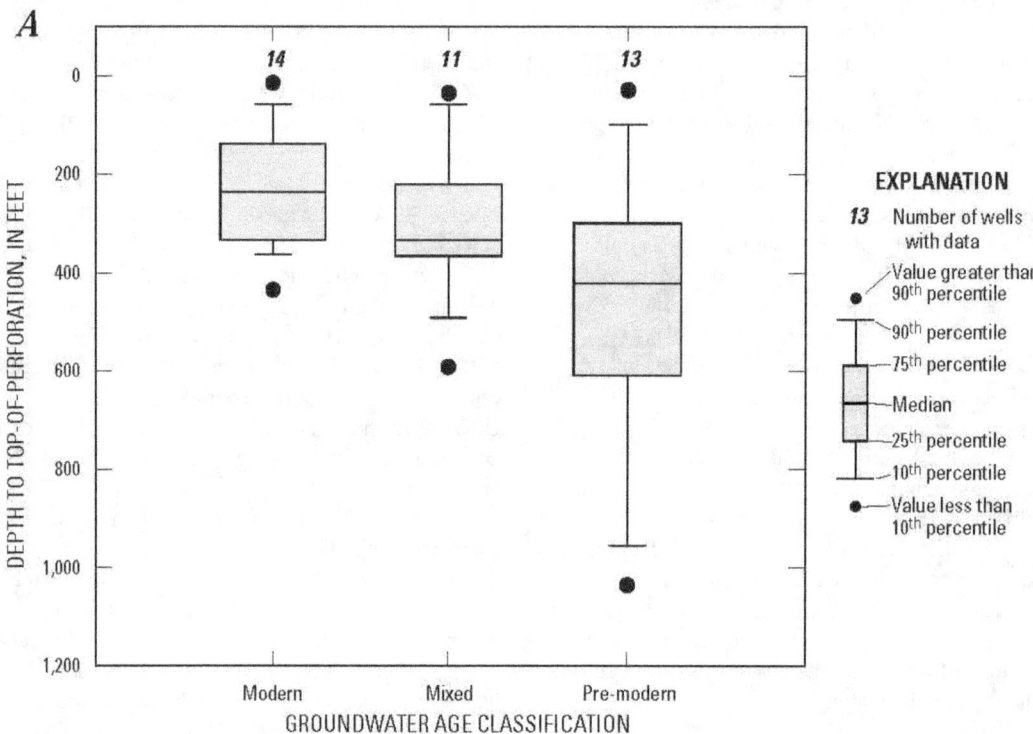

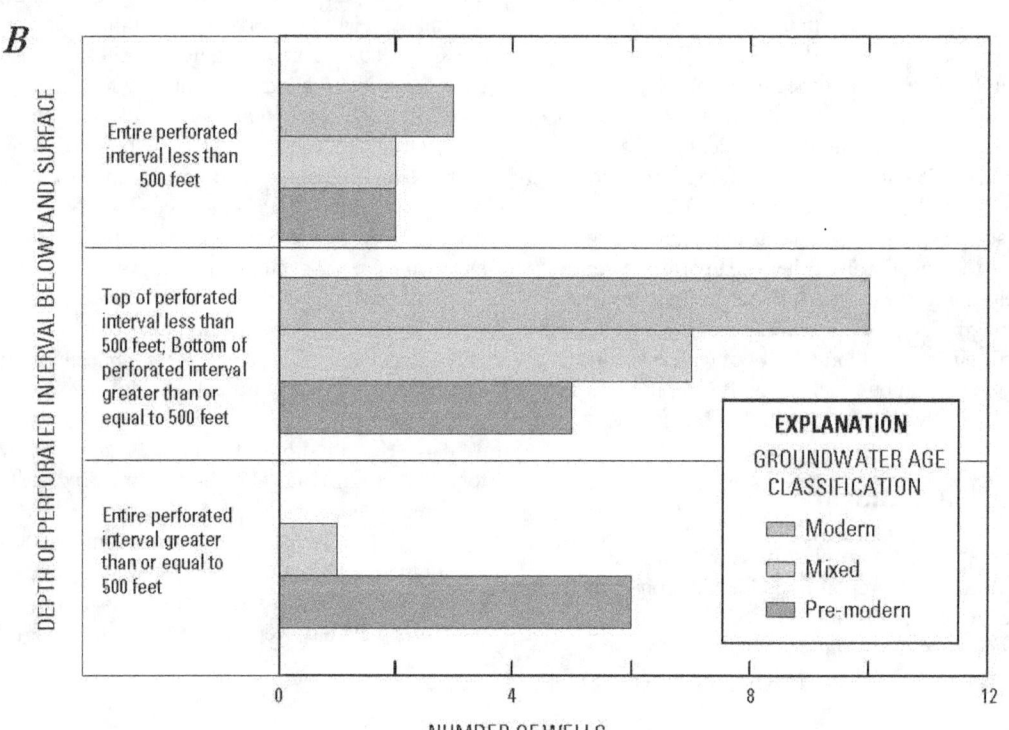

Figure 11.   Boxplots and bar charts showing the relation of groundwater age classification to (*A*) depth to top-of-perforations and (*B*) numbers of wells with each groundwater age class in each of the three depth categories, Kern County Subbasin study unit, California GAMA Priority Basin Project.

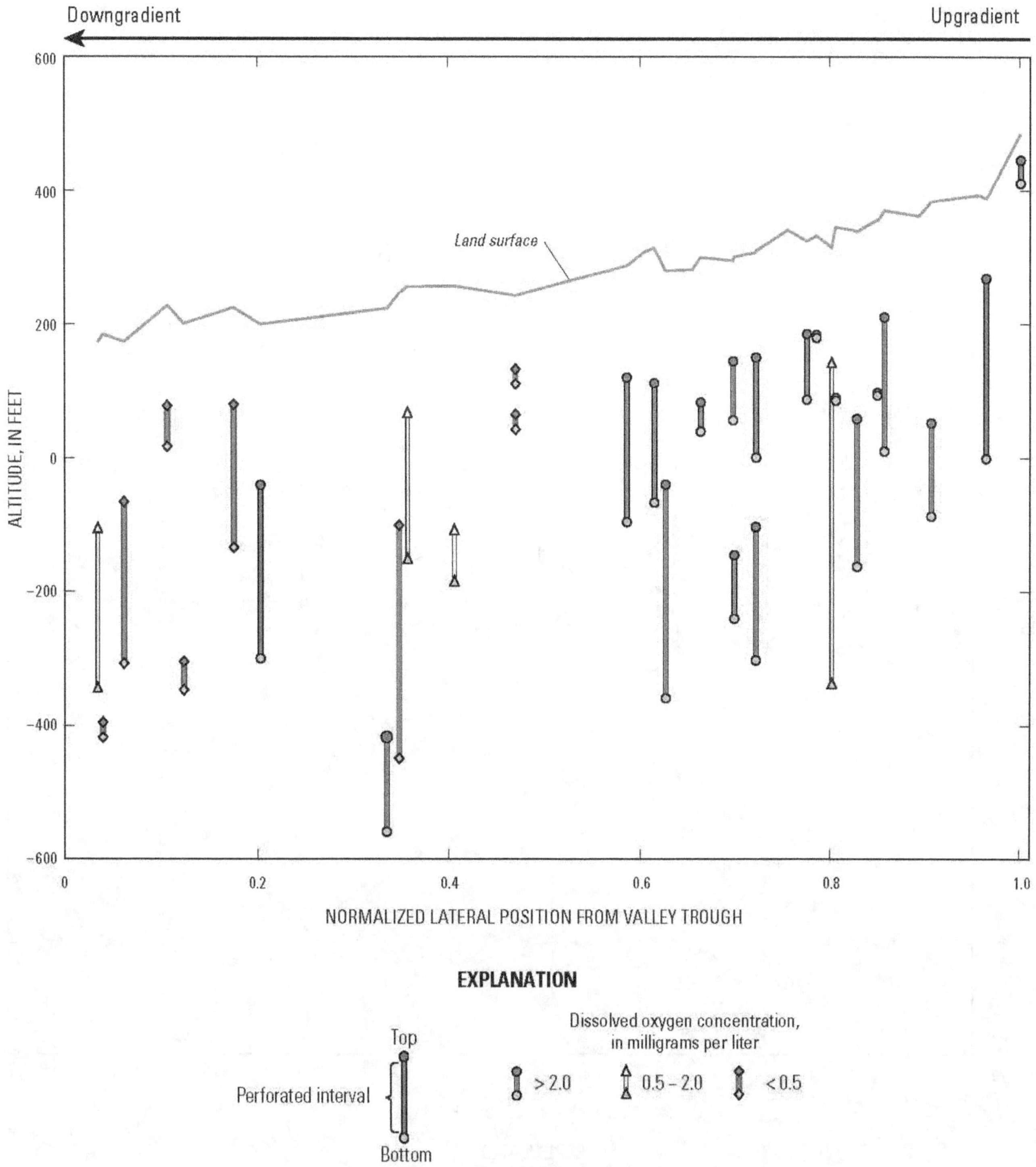

Downgradient
Upgradient
Land surface
ALTITUDE, IN FEET
600
400
200
0
-200
-400
-600
0
0.2
0.4
0.6
0.8
1.0
NORMALIZED LATERAL POSITION FROM VALLEY TROUGH
EXPLANATION
Top
Perforated interval
Bottom
Dissolved oxygen concentration, in milligrams per liter
>2.0
0.5 – 2.0
<0.5

Figure 12. Relation of oxidation-reduction category to lateral position and depths of perforated interval of wells in the Kings study area of the Southeast San Joaquin Valley study unit, California GAMA Priority Basin Project.

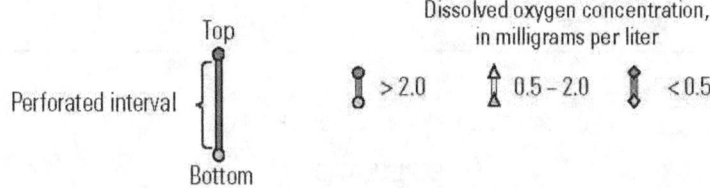

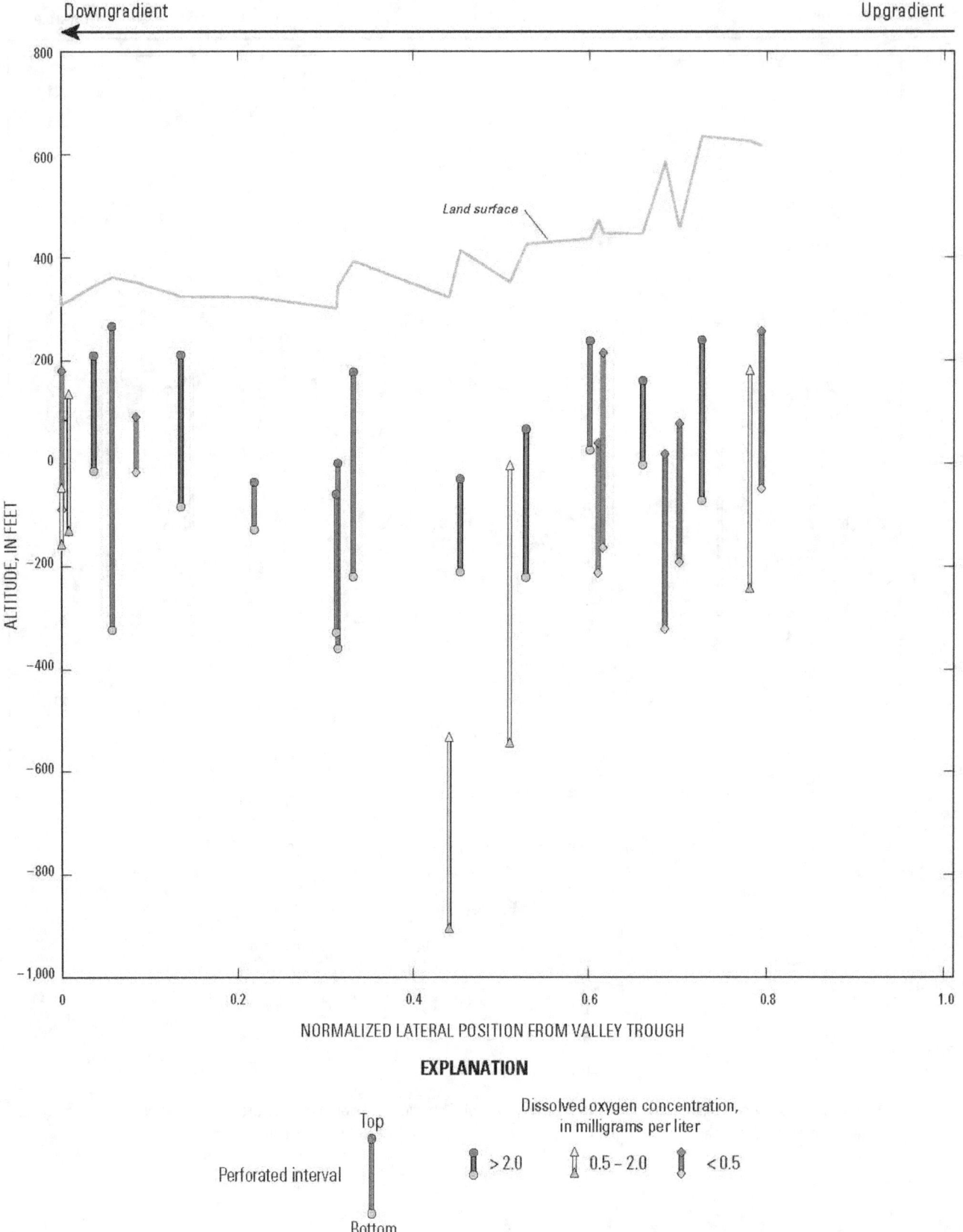

Figure 13.   Relation of oxidation-reduction category to lateral position and depth of perforated interval of wells located adjacent to or north of the Kern River, Kern County Subbasin study unit, California GAMA Priority Basin Project.

Other studies have shown that pH is correlated to several inorganic constituents in the San Joaquin Valley (Belitz and others, 2003; Welch and others, 2006; Izbicki and others, 2008; Landon and others, 2010a). pH ranged from 6.6 to 9.8 in 80 USGS- and CDPH-grid and USGS-understanding wells in the SESJ study unit (table D7; fig. 14A) with a median pH of 7.9. Median pH in the Kings study area was 7.7, in the Kaweah study area 8.1, in the Tule study area 8.0, and in the Tulare Lake study area 8.5. pH ranged from 5.5 to 9.6 in 54 USGS- and CDPH-grid and USGS-understanding wells in the KERN study unit (table D8; fig. 14A) with a median pH of 8.1. Data for pH also were available for the domestic wells sampled as part of the Domestic Well Project. The pH for the domestic wells was significantly lower than the pH for USGS- and CDPH-grid and USGS-understanding wells in either study unit (table 5; fig. 14A). The pH for domestic wells ranged from 6.3 to 8.4 in 143 domestic wells (table D9; fig. 14A) with a median pH of 7.2.

Correlations Between Explanatory Variables

Significant correlations between explanatory variables are important to identify because apparent correlations between an explanatory variable and a water-quality constituent could indicate relations between two explanatory variables and not between an explanatory variable and a water-quality constituent. Significant correlations using the Wilcoxon or Kruskal-Wallis statistical tests between categorical explanatory variables are given in table 5 for the southern San Joaquin Valley study units and domestic wells. Significant correlations using the Spearman's method between continuous explanatory variables are given in tables 6A, 6B, and 6C.

In addition to the relations of groundwater age and depth already discussed, well depth varied by study unit. Wells in the KERN study unit were deeper than wells in the SESJ study unit. In turn, the wells sampled for the Priority Basin Project were deeper than the wells sampled for the Domestic Well Project (table 5).

Well depth and depth to the top-of-perforations were positively correlated with urban land use in the KERN study unit (table 6A). In general, wells were deeper in urban areas than in agricultural land-use areas. Well depth and depth to top-of-perforations were not correlated with land use in the SESJ study unit or in the Domestic Well Project (tables 6B and 6C). Another study in the Central Eastside San Joaquin Valley (Landon and others, 2010a) also did not observe correlations of well depth with land use.

Groundwater age was related to other explanatory factors in ways that varied by study unit. In the KERN study unit, the percentage of agricultural land use was higher in wells with groundwater classified as pre-modern than in wells with groundwater classified as modern (table 5). This correlation was not observed for the SESJ study unit. In contrast, pH was significantly higher in wells with pre-modern age groundwater than in wells with modern age groundwater in the SESJ study unit but not in the KERN study unit. However, if the wells located on the western side of the valley trough in KERN are not included in the correlation, pH is higher in wells with pre-modern age groundwater than in wells with modern age groundwater (p=0.013). DO generally was higher in modern age groundwater than in pre-modern groundwater for both study units.

Correlations between lateral position and depth vary between study units. Grid wells in the SESJ study unit have a negative correlation with lateral position for well depth and depth to top-of-perforations (table 6B). This correlation indicates that wells generally are deeper in the distal part of the study unit than in the proximal part of the study unit. In contrast, lateral position is positively correlated with depth to top-of-perforations in the KERN study unit, implying that depth to top-of-perforations is greater near the boundary of the valley than in the center of the valley (table 6A).

Lateral position is correlated negatively with agricultural land use in both study units, and positively with urban land use in the KERN study unit and natural land use in the SESJ study unit. These correlations were expected because agricultural land use is more prevalent in the distal part of the study units, and urban land use in the KERN study unit and natural land use in the SESJ study unit are more prevalent along the eastern boundary (fig. 4).

Concentrations of DO were correlated significantly (positively) with lateral position in the SESJ study unit (table 6B), and negatively with well depth and depth to the top-of-perforations (fig. 12). Wells in the eastern (upgradient) part of the study unit have higher DO concentrations than wells in the western part. The correlation of DO with well depth may partially indicate the relation of DO with lateral position because the deep wells generally are in the western part of the SESJ study unit (table 6B). DO in the KERN study unit also was negatively correlated with depth to top-of-perforations. However, DO was not correlated significantly with normalized lateral position in the KERN study unit (table 6A). This was not unexpected because DO concentrations are low at the eastern boundary and near the valley trough of the study unit (fig. 13). Many of the wells in the eastern part of KERN have the deepest depth from land surface to the top-of-perforations.

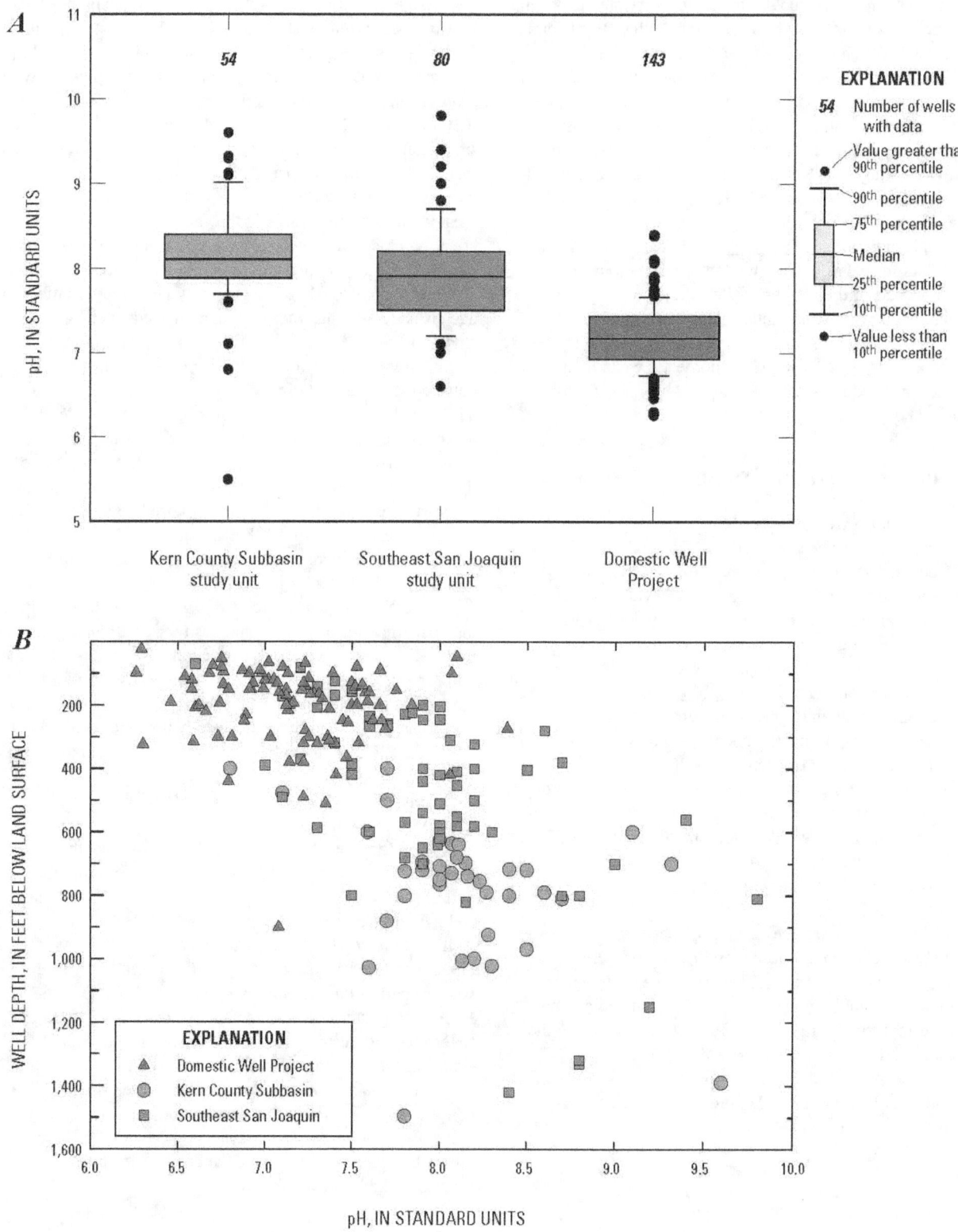

Figure 14. Relation of (*A*) pH and (*B*) well depth between the two southern San Joaquin Valley study units, California GAMA Priority Basin Project, and the domestic wells from the GAMA Domestic Well Project.

Table 6A. Results of nonparametric (Spearman's method) analysis of correlations in grid and understanding wells between selected potential explanatory factors, Kern County Subbasin study unit, California GAMA Priority Basin Project.

[Spearmans's rho (ρ) values and p-values are shown when correlations between selected potential explanatory factors are significant ($p < 0.05$). Number of septic tanks or cesspools in 500-meter radius around each well (U.S. Census Bureau, 1990). Number of formerly leaking underground fuel tanks within a Thiessen polygon in square kilometers, data from Geographic Information Management System GeoTracker (California Environmental Protection Agency, 2001). **Abbreviations**: ρ, Spearman's correlation statistic; p, significance level of Spearman's test based on a threshold value (α) of 0.05; mg/L, milligrams per liter; ns, not significant; <, less than]

Type of well analyzed	Explanatory factor	ρ :Spearman's correlation statistic/p: significance level						
		Depth of well	Depth to top-of-perforations	Normalized lateral position from valley trough	Dissolved oxygen (mg/L)	pH	Number of septic tanks or cesspools	Number of formerly leaking underground fuel tanks
Grid wells	Orchard or vineyard land use (percent)	ns	0.428 0.006	ns	ns	ns	ns	ns
	Agricultural land use (percent)	ns	ns	−0.323 0.027	ns	ns	−0.331 0.019	−0.383 0.006
	Natural land use (percent)	ns	ns	ns	ns	ns	ns	ns
	Urban land use (percent)	0.294 0.042	0.345 0.029	0.295 0.044	ns	ns	0.372 0.008	0.541 <0.001
Grid and understanding wells	Depth of well		0.565 <0.001	ns	ns	ns	ns	ns
	Depth to top-of-perforations			0.483 0.002	−0.354 0.025	ns	ns	ns
	Normalized lateral position from valley trough				ns	ns	ns	0.301 0.034
	Dissolved oxygen (mg/L)					ns	ns	ns
	pH						ns	ns
	Number of septic tanks or cesspools							0.681 <0.001

Table 6B.   Results of nonparametric (Spearman's method) analysis of correlations in grid and understanding wells between selected potential explanatory factors, Southeast San Joaquin Valley study unit, California GAMA Priority Basin Project.

[Spearmans's rho (ρ) values and p-values are shown when correlations between selected potential explanatory factors are significant ($p < 0.05$). Number of septic tanks or cesspools in 500-meter radius around each well (U.S. Census Bureau, 1990). Number of formerly leaking underground fuel tanks within a Thiessen polygon in square kilometers, data from Geographic Information Management System GeoTracker (California Environmental Protection Agency, 2001). **Abbreviations**: ρ, Spearman's correlation statistic, p, significance level of Spearman's test based on a threshold value (α) of 0.05; mg/L, milligrams per liter; ns, not significant; <, less than]

Type of well analyzed	Explanatory factor	ρ :Spearman's correlation statistic/p: significance level						
		Depth of well	Depth to top-of-perforations	Normalized lateral position from valley trough	Dissolved oxygen (mg/L)	pH	Number of septic tanks or cesspools	Number of formerly leaking underground fuel tanks
Grid wells	Orchard or vineyard land use (percent)	ns	ns	0.274 0.012	0.203 0.045	−0.342 0.005	0.244 0.015	ns
	Agricultural land use (percent)	ns	ns	−0.285 0.009	ns	ns	ns	−0.480 <0.001
	Natural land use (percent)	ns	ns	0.259 0.018	ns	ns	ns	0.205 0.042
	Urban land use (percent)	ns	ns	ns	ns	ns	0.231 0.022	0.479 <0.001
Grid and understanding wells	Depth of well		0.748 <0.001	−0.267 0.0154	−0.347 0.002	0.626 <0.001	ns	ns
	Depth to top-of-perforations			−0.423 <0.001	−0.389 <0.001	0.621 <0.001	−0.293 0.012	ns
	Normalized lateral position from valley trough				0.547 <0.001	−0.604 <0.001	0.434 <0.001	0.432 <0.001
	Dissolved oxygen (mg/L)					−0.320 0.009	0.244 0.0155	0.349 <0.001
	pH						ns	ns
	Number of septic tanks or cesspools							0.564 <0.001

Table 6C. Results of nonparametric (Spearman's method) analysis of correlations in domestic wells between selected potential explanatory factors, California GAMA Domestic Well Project.

[Spearmans's rho (ρ) values and p-values are shown when correlations between selected potential explanatory factors are significant (p < 0.05). Number of septic tanks or cesspools in 500-meter radius around each well (U.S. Census Bureau, 1990). **Abbreviations**: ρ, Spearman's correlation statistic; p, significance level of Spearman's test based on a threshold value (α) of 0.05; ns, not significant; <, less than]

Explanatory factor	ρ :Spearman's correlation statistic/ p: significance level			
	Depth of well	Normalized lateral position from valley trough	pH	Number of septic tanks or cesspools
Orchard or vineyard land use (percent)	−0.349 <0.001	0.296 <0.001	−0.200 0.017	0.316 <0.001
Agricultural land use (percent)	ns	−0.422 <0.001	0.205 0.014	ns
Natural land use (percent)	ns	0.515 <0.001	−0.289 <0.001	ns
Urban land use (percent)	ns	ns	ns	0.401 <0.001
Depth of well		ns	0.223 0.029	−0.321 0.001
Normalized lateral position from valley trough			−0.478 <0.001	ns
pH				ns

pH was positively correlated with well depth and depth to the top of the perforations in the SESJ study unit; pH was higher for deep wells than for shallow wells (table 6B; fig. 14B). pH was negatively correlated to lateral position and DO; pH was higher in the distal part of the SESJ study unit where DO concentrations are lower. The correlations of pH with lateral position and DO may indicate the relation of pH to well depth. pH also was significantly lower in wells with modern age groundwater than in wells with pre-modern age groundwater but not in wells with mixed age groundwater

(table 5). Similar to other explanatory variables, correlations with pH vary between the two study units. pH was not correlated with any explanatory variables in the KERN study unit (table 6A). However, if the wells on the western side of the valley trough are not included in the correlations, pH is positively correlated with well depth (rho=0.569, p=0.001) and depth to top-of-perforations (rho=0.506; p=0.010). Implications of correlations between explanatory factors are discussed later in the report as part of analysis of factors affecting individual constituents.

Status and Understanding of Water Quality

The *status assessment* was designed to identify the constituents or classes of constituents most likely to be water-quality concerns because of high relative-concentrations or prevalence. The assessment applies only to constituents with regulatory (MCL and AL) or non-regulatory (HAL, RSD5-US, or NL) human-health benchmarks or aesthetic benchmarks (SMCL) established by the USEPA or the CDPH (U.S. Environmental Protection Agency, 2008a, 2008b; California Department of Public Health, 2008a). The spatially distributed, randomized approach to well selection and data analysis yields a view of groundwater quality in which all areas of the primary aquifers are weighted equally.

The *understanding assessment* was designed to help answer the question of why specific constituents are, or are not, detected in groundwater. The understanding assessment addresses a subset of the constituents discussed in the status assessment and is based on statistical correlations between water quality and a finite set of potential explanatory factors. This assessment may improve our understanding of how human and natural sources of contaminants affect groundwater quality in the southern San Joaquin Valley; however, it was not designed to identify specific sources of constituents to specific wells.

In USGS-grid wells, less than one-third of organic and special-interest constituents analyzed for were detected (68 of 226). Human-health benchmarks are established for about two-thirds of the organic and special-interest constituents (46 of 68) detected (table 7). Twenty-eight VOCs, including gasoline oxygenates, were detected; human-health benchmarks established for all but five VOCs. Human-health benchmarks were established for all five fumigants detected. Thirty-three pesticides were detected; human-health benchmarks were established for 18 of the 33 pesticides (Burton and Belitz, 2008; Shelton and others, 2008). Eight of the detected pesticides (de-ethylatrazine, de-ethyl-deisopropylatrazine, de-isopropylatrazine, 1-napthol, 3,4-dichloroanaline, 3,5-dichloroaniline, desulfinyl fipronil, and fipronil sulfide) with no benchmarks are pesticide degradates; human-health benchmarks are established for three (atrazine, carbaryl, and diuron) of the five parent compounds

of these degradates. Human-health benchmarks have not been established for the parent compounds (fipronil and iprodione) of the remaining degradates that were detected. Human-health benchmarks are established for both constituents of special interest detected. Thus, the organic and special-interest constituents that are regulated include most of these constituents that were detected in groundwater in the southern San Joaquin Valley study units.

In contrast to organic constituents, inorganic constituents nearly always were detected (52 of 53, table 7) in USGS- and CDPH-grid wells. Human-health or aesthetic benchmarks have been established for almost three-quarters of inorganic constituents detected (38 of 52). Most of the constituents without benchmarks are major or minor ions that are naturally present in nearly all groundwater.

The maximum relative-concentration for each constituent with a water-quality benchmark in grid wells is shown in figure 15. In the SESJ study unit, nine inorganic constituents (including radioactive constituents) were detected at high relative-concentrations in one or more grid wells, and six additional inorganic constituents were detected at moderate relative-concentrations (fig. 15A). In contrast, three of the organic and special-interest constituents were detected at high relative-concentrations in grid wells, and three additional organic and special-interest constituents were detected at moderate relative-concentrations. In the KERN study unit, ten inorganic constituents were detected at high relative-concentrations in one or more grid wells, and seven additional inorganic constituents were detected at moderate relative-concentrations (fig. 15B). Only one organic constituent was detected at high relative-concentrations, and six additional organic and special-interest constituents were detected at moderate relative-concentrations.

Aquifer-scale proportions were calculated for each inorganic and organic constituent detected at high or at moderate relative-concentrations and for each organic and special-interest constituent detected in more than 10 percent of the grid wells (tables 8 and 9). Spatially weighted high aquifer-scale proportions were within the 90 percent confidence intervals for their respective grid-based aquifer high proportions for all constituents (see tables 8 and 9) except for unadjusted gross alpha radioactivity from KERN, providing evidence that the grid-based approach yields statistically equivalent results to the spatially weighted approach.

Table 7. Number of constituents analyzed and detected in USGS-grid wells by human-health-based or aesthetic benchmark and constituent type in the two southern San Joaquin Valley study units, California GAMA Priority Basin Project.

[Regulatory human-health benchmarks include U.S. Environmental Protection Agency (USEPA) and California Department of Public Health (CDPH) maximum contaminant levels. Non-regulatory human-health benchmarks include USEPA lifetime health advisory levels and risk-specific dose level at 10^{-5} lifetime cancer risk and CDPH notification level. **Abbreviations**: SESJ, Southeast San Joaquin Valley study unit; KERN, Kern County Subbasin study unit; VOCs, volatile organic compounds; NWQL, USGS National Water Quality Laboratory; HHB, human-health-based benchmark; SMCL, USEPA or CDPH Secondary Maximum Contaminant Level (aesthetic based); USGS, U.S. Geological Survey; GAMA, Groundwater Ambient Monitoring and Assessment Program]

Benchmark type	Number of constituents analyzed	Number of constituents detected		
		SESJ	KERN	SESJ and KERN
VOCs + Gasoline oxygenates (excluding fumigants)				
Regulatory HHB	29	19	11	19
Non-regulatory HHB	22	4	3	4
None	27	5	0	5
Total	78	28	14	28
Fumigants				
Regulatory HHB	4	3	3	4
Non-regulatory HHB	4	1	1	1
None	2	0	0	0
Total	10	4	4	5
Pesticides and degradates (NWQL Schedule 2033)				
Regulatory HHB	5	3	2	3
Non-regulatory HHB	18	5	6	7
None	59	9	4	12
Total	82	17	12	22
Polar pesticides and degradates (NWQL Schedule 2060)				
Regulatory HHB	7	3	1	3
Non-regulatory HHB	9	2	3	3
None	38	5	2	5
Total	54	10	6	11
Special interest				
Regulatory HHB	1	1	1	1
Non-regulatory HHB	1	1	0	1
Total	2	2	1	2
Sum of inorganic and radioactive constituents				
Regulatory HHB	25	24	24	24
Non-regulatory HHB	9	8	8	9
Aesthetic - SMCL	5	5	5	5
None	14	14	14	14
Total	53	51	51	52

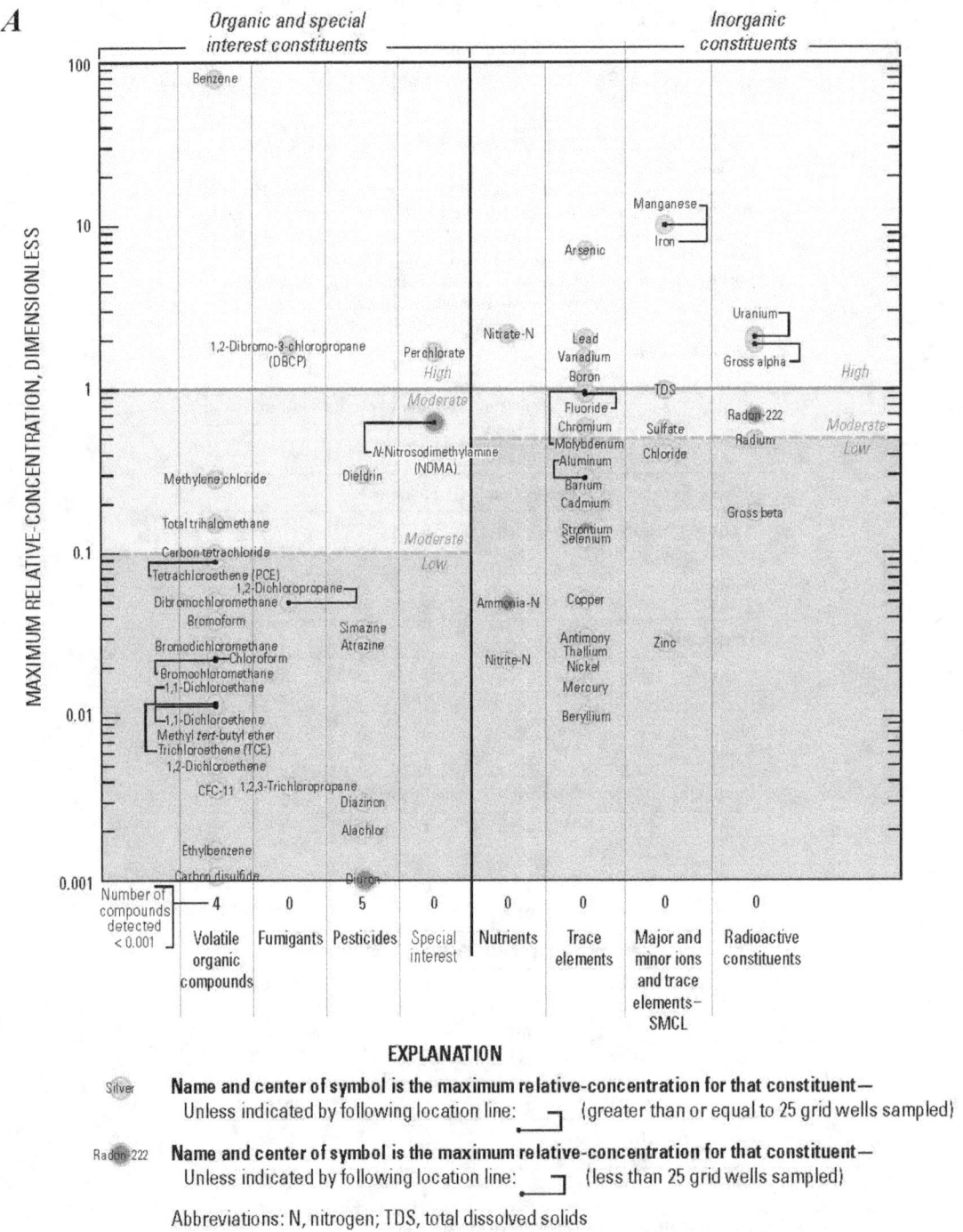

Figure 15. Graphs showing maximum relative-concentration in USGS- and CDPH-grid wells for constituents detected by type of constituent in the (*A*) Southeast San Joaquin Valley study unit, and (*B*) Kern County Subbasin study unit, California GAMA Priority Basin Project.

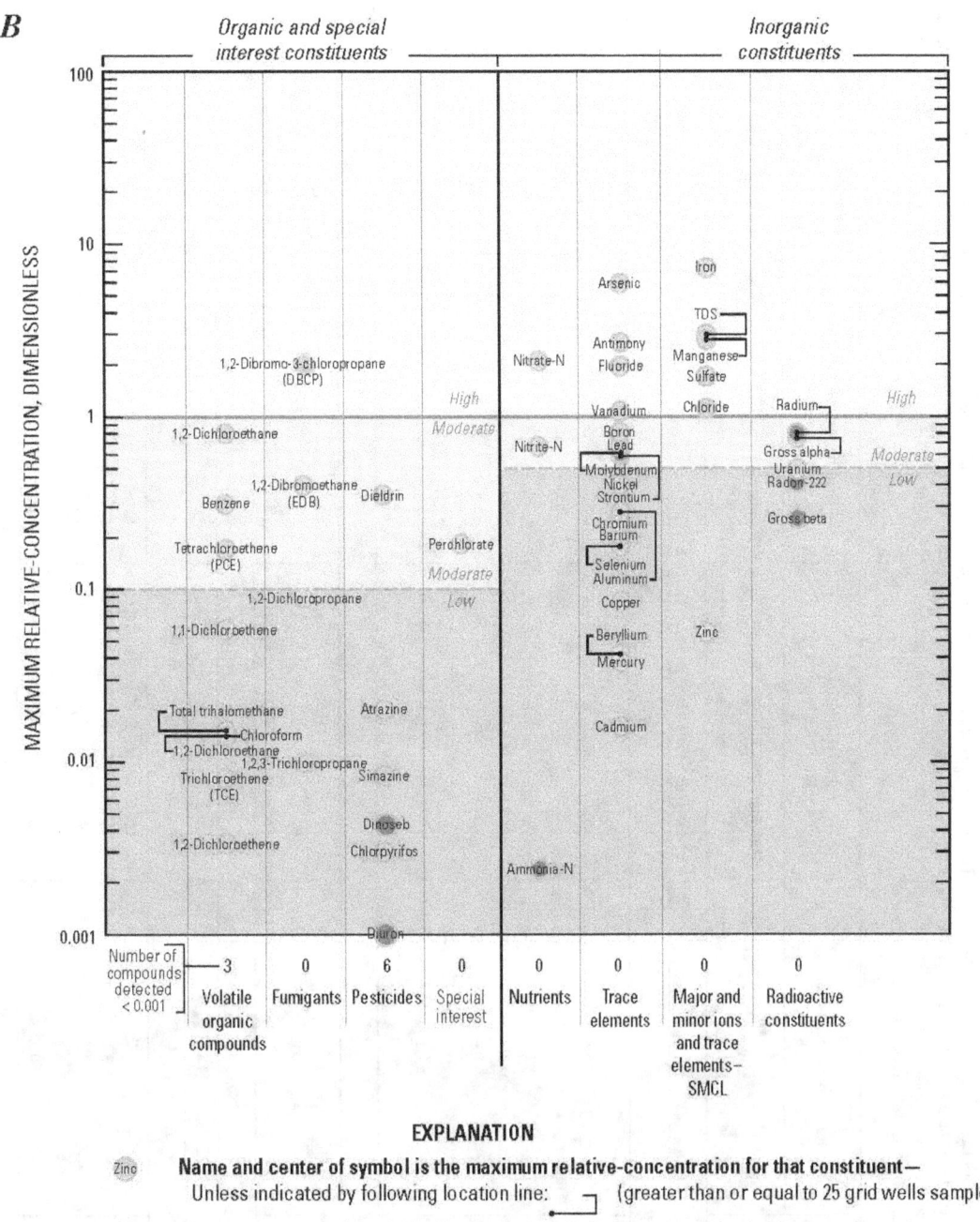

EXPLANATION

Zinc   **Name and center of symbol is the maximum relative-concentration for that constituent—**
Unless indicated by following location line: (greater than or equal to 25 grid wells sampled)

Radon-222   **Name and center of symbol is the maximum relative-concentration for that constituent—**
Unless indicated by following location line: (less than 25 grid wells sampled)

Abbreviations: N, nitrogen; TDS, total dissolved solids

Figure 15.—Continued

Table 8. Raw detection frequencies and aquifer-scale proportions using spatially weighted and grid-based methods for inorganic constituents detected at concentrations greater than water-quality benchmarks from data available during the most recent 3 years (January 1, 2003, through December 31, 2005) in the California Department of Public Health (CDPH) database, or detected at high or at moderate relative-concentrations in samples collected from USGS-grid wells (October 2005–March 2006) in the two southern San Joaquin Valley study units, California GAMA Priority Basin Project.

[Grid-based aquifer-scale proportions for inorganic constituents are based on samples collected by the U.S. Geological Survey from grid wells during October 2005–March 2006 and supplemented with additional data from selected wells in the CDPH data base (January 1, 2003, through December 31, 2005). High, concentrations greater than water-quality benchmark; moderate, concentrations less than or equal to benchmark but greater than 0.5 of benchmark; low, concentrations less than 0.5 of benchmark. Constituents are ordered by class and high aquifer-scale proportion. **Abbreviation:** nc, not collected; USGS, U.S. Geological Survey; GAMA, Groundwater Ambient Monitoring and Assessment Program]

Constituent	Raw detection frequency[1]			Spatially weighted aquifer-scale proportion[1]			Grid-based aquifer-scale proportion			90 percent confidence interval for grid-based high aquifer-scale proportion[2]	
	Number of wells	Moderate values (percent)	High values (percent)	Number of cells	Moderate values (percent)	High values (percent)	Number of cells	Moderate values (percent)	High values (percent)	Lower limit (percent)	Upper limit (percent)
Southeast San Joaquin Valley study unit											
Trace and minor elements											
Arsenic	752	5.5	9.2	76	11	21	73	12	19	13	28
Boron	290	2.1	1.0	57	4.5	3.9	46	6.5	6.5	2.4	15
Vanadium	333	23	4.5	60	27	5.6	49	25	6.1	2.2	14
Lead	741	1.2	0.3	73	1.1	1.5	69	0.0	1.4	0.3	5.5
Fluoride	723	3.3	0.0	71	5.5	0.0	67	10	0.0	0.0	2.8
Molybdenum	nc	nc	nc	nc	nc	nc	28	3.6	0.0	0.0	4.7
Chromium	693	0.4	0.0	75	0.6	0.0	71	1.4	0.0	0.0	1.9
Aluminum	743	1.1	0.4	75	1.7	0.5	72	0.0	0.0	0.0	1.9
Barium	742	0.0	0.1	75	0.0	0.3	72	0.0	0.0	0.0	1.9
Radioactive constituents											
Uranium[3]	156	10	3.2	51	11	6.9	40	5.0	5.0	1.4	13
Gross alpha radioactivity	628	10	3.2	62	11	4.7	53	7.5	3.8	1.1	10
Radon-222	nc	nc	nc	nc	nc	nc	9	11	0.0	0.0	14
Nutrients											
Nitrate plus nitrite	975	18	5.9	81	17	9.0	79	16	6.3	2.9	12
Inorganics with aesthetic benchmarks (SMCLs)											
Manganese	682	1.0	3.8	65	2.7	8.5	61	1.6	4.9	1.8	11
Iron	684	3.8	4.5	65	6.5	8.4	61	4.9	3.3	0.9	8.8
Total dissolved solids (TDS)	679	3.5	0.3	65	7.5	0.4	84	13	0.0	0.0	1.6
Sulfate	680	0.3	0.0	65	0.8	0.0	61	1.7	0.0	0.0	2.2
Chloride	679	0.1	0.1	65	0.2	0.3	61	0.0	0.0	0.0	2.2

Table 8. Raw detection frequencies and aquifer-scale proportions using spatially weighted and grid-based methods for inorganic constituents detected at concentrations greater than water-quality benchmarks from data available during the most recent 3 years (January 1, 2003, through December 31, 2005) in the California Department of Public Health (CDPH) database, or detected at high or at moderate relative-concentrations in samples collected from USGS-grid wells (October 2005–March 2006) in the two southern San Joaquin Valley study units, California GAMA Priority Basin Project.—Continued

[Grid-based aquifer-scale proportions for inorganic constituents are based on samples collected by the U.S. Geological Survey from grid wells during October 2005–March 2006 and supplemented with additional data from selected wells in the CDPH data base (January 1, 2003, through December 31, 2005). High, concentrations greater than water-quality benchmark; moderate, concentrations less than or equal to benchmark but greater than 0.5 of benchmark; low, concentrations less than 0.5 of benchmark. Constituents are ordered by class and high aquifer-scale proportion. **Abbreviation:** nc, not collected; USGS, U.S. Geological Survey; GAMA, Groundwater Ambient Monitoring and Assessment Program]

Constituent	Raw detection frequency[1]			Spatially weighted aquifer-scale proportion[1]			Grid-based aquifer-scale proportion			90 percent confidence interval for grid-based high aquifer-scale proportion[2]	
	Number of wells	Moderate values (percent)	High values (percent)	Number of cells	Moderate values (percent)	High values (percent)	Number of cells	Moderate values (percent)	High values (percent)	Lower limit (percent)	Upper limit (percent)
Kern County Subbasin study unit											
Trace and minor elements											
Arsenic	353	19	12	57	21	19	56	21	20	12	29
Antimony	346	1	2.0	57	1.8	4.7	56	1.8	3.6	1.0	9.5
Fluoride	349	0.6	0.6	58	2.0	3.4	57	1.8	3.5	1.0	9.4
Vanadium	188	3.2	2.1	57	0.0	1.1	29	0.0	3.4	2.0	18
Boron	161	1.9	1.2	36	4.9	2.1	32	9.4	0.0	0.0	4.1
Molybdenum	20	5.0	0.0	17	5.9	0.0	14	7.1	0.0	0.0	9.1
Strontium	nc	nc	nc	nc	nc	nc	14	7.1	0.0	0.0	9.1
Lead	335	0.9	1.8	55	0.7	1.2	53	1.9	0.0	0.0	2.5
Thallium	346	0.0	0.9	57	0.0	1.1	56	0.0	0.0	0.0	2.4
Selenium	346	0.0	0.3	57	0.0	0.3	56	0.0	0.0	0.0	2.4
Radioactive constituents											
Gross alpha radioactivity	257	8.2	3.1	42	12	4.0	38	13	0.0	0.0	3.5
Radium	123	0.8	1.6	24	4	4.5	18	5.6	0.0	0.0	10
Uranium	83	4.8	4.8	28	2.5	6.1	14	0.0	0.0	0.0	9.1
Nutrients											
Nitrate plus nitrite	391	13	4.9	67	15	5.6	67	12	4.5	1.6	10
Nitrite	360	0.3	0.0	61	0.8	0.0	60	1.7	0.0	0.0	2.2
Inorganics with aesthetic benchmarks (SMCLs)											
Total dissolved solids (TDS)	336	9.2	6.0	58	16	16	58	17	14	7.7	23
Sulfate	316	3.5	1.9	48	6.1	8.4	48	6.2	8.3	3.5	17
Chloride	316	2.5	0.3	48	3.8	1.0	48	4.2	2.1	0.4	7.9
Iron	329	3.0	6.1	53	5.0	9.5	53	3.8	9.4	4.4	18
Manganese	327	1.2	3.4	53	1.0	5.4	53	1.9	5.7	2.1	13

[1] Based on most recent analysis for each CDPH well during January 1, 2003–December 31, 2005, combined with GAMA grid-based data.

[2] Based on the Jeffreys interval for the binomial distribution (Brown and others, 2001).

[3] Gross alpha activities were not adjusted for uranium activity. The MCL-US for gross alpha activity applies to adjusted gross alpha activity.

Table 9. Raw detection frequencies and aquifer-scale proportions using spatially weighted and grid-based methods for organic constituents and constituents of special interest detected at concentrations greater than water-quality benchmarks from data available during the most recent 3 years (January 1, 2003, through December 31, 2005) in the California Department of Public Health (CDPH) database, or detected at high or at moderate relative-concentrations, or detected at detection frequencies greater than 10 percent in samples collected from USGS-grid wells (October 2005–March 2006) in the two southern San Joaquin Valley study units, California GAMA Priority Basin Project.

[Grid-based aquifer-scale proportions for organic constituents are based on samples collected by the U.S. Geological Survey from grid wells during October 2005–March 2006. High, concentrations greater than water-quality benchmark; moderate, concentrations less than or equal to benchmark but greater than 0.1 of benchmark; low, concentrations less than 0.1 of benchmark. Constituents are ordered by class and high aquifer-scale proportion. **Abbreviation:** nc, not collected]

Constituent	Raw detection frequency[1]			Spatially weighted aquifer-scale proportion[1]			Grid-based aquifer-scale proportion			90 percent confidence interval for grid-based high aquifer-scale proportion[2]	
	Number of wells	Moderate values (percent)	High values (percent)	Number of cells	Moderate values (percent)	High values (percent)	Number of cells	Moderate values (percent)	High values (percent)	Lower limit (percent)	Upper limit (percent)
Southeast San Joaquin Valley study unit											
Trihalomethanes (THMs)[3]											
Total THMs	773	0.8	0.0	83	2.5	0.0	83	1.2	0.0	0.0	1.6
Chloroform	773	0.1	0.0	85	0.1	0.0	83	0.0	0.0	0.0	1.6
Solvents											
Dichloromethane (methylene chloride)	771	0.3	0.0	85	1.3	0.0	83	1.2	0.0	0.0	1.6
Carbon tetrachloride	771	0.0	0.4	85	0.0	0.3	83	0.0	0.0	0.0	1.6
Tetrachloroethene (PCE)	771	2.7	0.4	85	0.8	0.3	83	0.0	0.0	0.0	1.6
Trichloroethene (TCE)	772	0.3	0.5	85	0.0	0.1	83	0.0	0.0	0.0	1.6
Other volatile organic compounds (VOCs)											
Benzene	773	0.0	0.4	85	0.0	1.2	83	0.0	1.2	0.2	4.6
Fumigants											
1,2-Dibromo-3-chloropropane (DBCP)	835	15	5.6	84	12	3.8	83	9.6	3.6	1.3	8.2
1,2-Dibromoethane (EDB)	825	0.1	0.2	84	0.1	0.04	83	0.0	0.0	0.0	1.6
Pesticides											
Dieldrin	272	0.7	0.0	83	0.7	0.0	83	2.4	0.0	0.0	1.6
Atrazine	765	0.0	0.0	84	0.0	0.0	83	0.0	0.0	0.0	1.6
Simazine	765	0.0	0.0	84	0.0	0.0	83	0.0	0.0	0.0	1.6
Bromacil	693	0.0	0.0	65	0.0	0.0	28	0.0	0.0	0.0	4.7
Constituents of special interest											
Perchlorate	297	8.0	0.7	84	13	1.2	83	18	1.2	0.2	4.6
N-Nitrosodimethylamine	nc	nc	nc	nc	nc	nc	29	3.4	0.0	0.0	4.5

Table 9. Raw detection frequencies and aquifer-scale proportions using spatially weighted and grid-based methods for organic constituents and constituents of special interest detected at concentrations greater than water-quality benchmarks from data available during the most recent 3 years (January 1, 2003, through December 31, 2005) in the California Department of Public Health (CDPH) database, or detected at high or at moderate relative-concentrations, or detected at detection frequencies greater than 10 percent in samples collected from USGS-grid wells (October 2005–March 2006) in the two southern San Joaquin Valley study units, California GAMA Priority Basin Project.—Continued

[Grid-based aquifer-scale proportions for organic constituents are based on samples collected by the U.S. Geological Survey from grid wells during October 2005–March 2006. High, concentrations greater than water-quality benchmark; moderate, concentrations less than or equal to benchmark but greater than 0.1 of benchmark; low, concentrations less than 0.1 of benchmark. Constituents are ordered by class and high aquifer-scale proportion. Abbreviation: nc, not collected]

Constituent	Raw detection frequency[1]			Spatially weighted aquifer-scale proportion[1]			Grid-based aquifer-scale proportion			90 percent confidence interval for grid-based high aquifer-scale proportion[2]	
	Number of wells	Moderate values (percent)	High values (percent)	Number of cells	Moderate values (percent)	High values (percent)	Number of cells	Moderate values (percent)	High values (percent)	Lower limit (percent)	Upper limit (percent)
Kern County Subbasin study unit											
Trihalomethanes (THMs)[3]											
Total THMs	355	1.1	0.0	60	4.2	0.0	47	0.0	0.0	0.0	2.8
Chloroform	355	0.3	0.0	60	1.7	0.0	47	0.0	0.0	0.0	2.8
Solvents											
Tetrachloroethene (PCE)	353	3.7	0.0	60	1.3	0.0	47	2.1	0.0	0.0	2.8
1,2-Dichloroethane (1,2-DCA)	352	0.3	0.0	60	0.8	0.0	47	2.1	0.0	0.0	2.8
Carbon tetrachloride	352	0.0	0.3	60	0.0	0.1	47	0.0	0.0	0.0	2.8
Trichloroethene (TCE)	353	0.6	0.3	60	0.0	0.04	47	0.0	0.0	0.0	2.8
Other volatile organic compounds (VOCs)											
Benzene	354	0.3	0.0	60	1.7	0.0	47	2.1	0.0	0.0	2.8
Fumigants											
1,2-Dibromo-3-chloropropane (DBCP)	378	4.8	1.1	61	5.0	1.2	47	2.1	2.1	0.4	8.0
1,2-Dibromoethane (EDB)	378	2.1	0.0	61	3.4	0.0	47	2.1	0.0	0.0	2.8
1,2-Dichloropropane (1,2-DCP)	352	0.9	0.0	60	1.2	0.0	47	0.0	0.0	0.0	2.8
1,2,3-Trichloropropane (1,2,3-TCP)	327	0.0	0.0	59	0.0	0.0	47	0.0	0.0	0.0	2.8
Pesticides											
Dieldrin	97	2.1	0.0	49	1.2	0.0	46	2.2	0.0	0.0	2.9
Atrazine	328	0.0	0.0	61	0.0	0.0	46	0.0	0.0	0.0	2.9
Simazine	328	0.0	0.0	61	0.0	0.0	46	0.0	0.0	0.0	2.9
Dinoseb	53	0.0	0.0	25	0.0	0.0	14	0.0	0.0	0.0	9.1
Constituents of special interest											
Perchlorate	149	2.0	0.0	49	5.1	0.0	47	6.4	0.0	0.0	2.8

[1] Based on most recent analysis for each CDPH well during January 1, 2003–December 31, 2005, combined with GAMA grid-based data.

[2] Based on the Jeffreys interval for the binomial distribution (Brown and others, 2001).

[3] The MCL-US (U.S. Environmental Protection Agency maximum contaminant level) for trihalomethanes is the sum of chloroform, bromoform, bromodichloromethane, and dibromochloromethane.

Inorganic Constituents

Inorganic constituents generally occur naturally in groundwater, although these concentrations may be affected by human as well as natural factors. In the study units, inorganic constituents with human-health benchmarks that were detected at high relative-concentrations in more than 2 percent of the primary aquifers were arsenic, boron, vanadium, nitrate, uranium, and gross alpha radioactivity. Inorganic constituents with aesthetic benchmarks that were detected at high relative-concentrations in the study units were iron and manganese. Additional constituents with human-health benchmarks—antimony, radium, and fluoride—and constituents with aesthetic benchmarks—total dissolved solids (TDS), sulfate, and chloride—were detected at high relative-concentrations in more than 2 percent of the primary aquifer in the KERN study unit but not in SESJ. Lead, aluminum, barium, thallium, and selenium were detected at high relative-concentrations in one or both study units in less than 2 percent of the primary aquifers (table 8). All detections are in untreated groundwater samples from the primary aquifers and not from drinking water, which is frequently treated before it is delivered to consumers by water purveyors.

Inorganic constituents in the SESJ study units with human-health benchmarks, as a group (trace elements and minor ions, uranium and radioactive constituents, and nutrients), were detected at high relative-concentrations in 30 percent, at moderate relative-concentrations in 30 percent, and at low relative-concentrations or not detected in 39 percent of the primary aquifers (table 10). Inorganic constituents with aesthetic benchmarks, as a group, were detected at high relative-concentrations in 6.6 percent, at moderate relative-concentrations in 13 percent, and at low relative-concentrations or not detected in 81 percent of the primary aquifers. In contrast, inorganic constituents in the KERN study unit with human-health benchmarks, as a group (trace elements and minor ions, uranium and radioactive constituents, and nutrients), were detected at high relative-concentrations in 23 percent, at moderate relative-concentrations in 29 percent, and at low relative-concentrations or not detected in 48 percent of the primary aquifers (table 10). Inorganic constituents with aesthetic benchmarks, as a group, were detected at high relative-concentrations in 22 percent, at moderate relative-concentrations in 17 percent, and at low relative-concentrations or not detected in 61 percent of the primary aquifers (table 10).

Trace Elements and Minor Ions

In the study units, the aquifer-scale proportions of one or more constituents for trace elements and minor ions with human-health benchmarks, as a class, were high. Trace elements in the SESJ study unit were detected at high relative-concentrations in 24 percent, at moderate relative-concentrations in 26 percent, and at low relative-concentrations or not detected in 50 percent of the primary aquifers (table 10). The aquifer-scale proportions for trace elements detected in the KERN study unit were similar to those detected in the SESJ study unit; trace elements were detected at high relative-concentrations in 20 percent, at moderate relative-concentrations in 27 percent, and at low relative-concentrations or not detected in 54 percent of the primary aquifers (table 10). Only constituents detected at high relative-concentrations in more than 2 percent of the primary aquifers in the study units are discussed further in this report.

Arsenic

Arsenic is a naturally occurring semi-metallic trace element. The most common source of arsenic is from aquifer materials in the southern San Joaquin Valley including dissolution of arsenic-rich minerals such as arsenopyrite, a common constituent of shales, and apatite, a common constituent of phosphorites. Anthropogenic sources of arsenic are from uses, for example, as a wood preservative, in paints and dyes, in drugs, and in the mining of copper and gold (Welch and others, 2000).

Arsenic was detected at high relative-concentrations in 19 and 20 percent of the primary aquifers in the SESJ and KERN study units, respectively (table 8). Arsenic was detected at high relative-concentrations in USGS- and CDPH-grid wells in the Kings, Tule, and Tulare Lake study areas in the SESJ study unit and in the KERN study unit (fig.16A). High relative-concentrations of arsenic also were detected in some CDPH-other wells in the Kaweah study area (fig. 17A). Most arsenic detections at high relative-concentrations were in the western part of the SESJ study unit. There also were several arsenic detections of high relative-concentrations near the city of Delano and to the south of the city of Bakersfield.

Table 10. Aquifer-scale proportions for inorganic constituent classes for the two southern San Joaquin Valley study units, California GAMA Priority Basin Project.

[Aquifer-scale proportions are given in percentage of area of the primary aquifer. All values greater than 10 percent are rounded to the nearest 1 percent, values less than 10 percent are rounded to the nearest 0.1 percent, values may not add up to 100 percent because of rounding. **Abbreviations**: SMCL, secondary maximum contaminant level; SO_4, sulfate; Cl, chloride; TDS, total dissolved solids]

Constituent	Aquifer-scale proportion (percent)		
	High	Moderate	Low or not detected
Southeast San Joaquin Valley study unit			
Inorganics with human-health benchmark			
Trace elements and minor ions	24	26	50
Uranium and radioactive constituents	6.9	6.9	86
Nutrients	6.3	19	75
Any inorganic with human-health benchmarks	30	30	39
Inorganics with aesthetic benchmark (SMCLs)			
Major ions (TDS, SO_4, Cl)	0.4	13	87
Manganese and (or) iron	6.6	4.9	88
Any inorganic with an SMCL	6.6	13	81
Kern County Subbasin study unit			
Inorganics with human-health benchmark			
Trace elements and minor ions	20	27	54
Uranium and radioactive constituents	6.1	13	81
Nutrients	4.5	13	82
Any inorganic with human-health benchmarks	23	29	48
Inorganics with aesthetic benchmark (SMCLs)			
Major ions (TDS, SO_4, Cl)	14	17	69
Manganese and (or) iron	13	5.7	82
Any inorganic with an SMCL	22	17	61

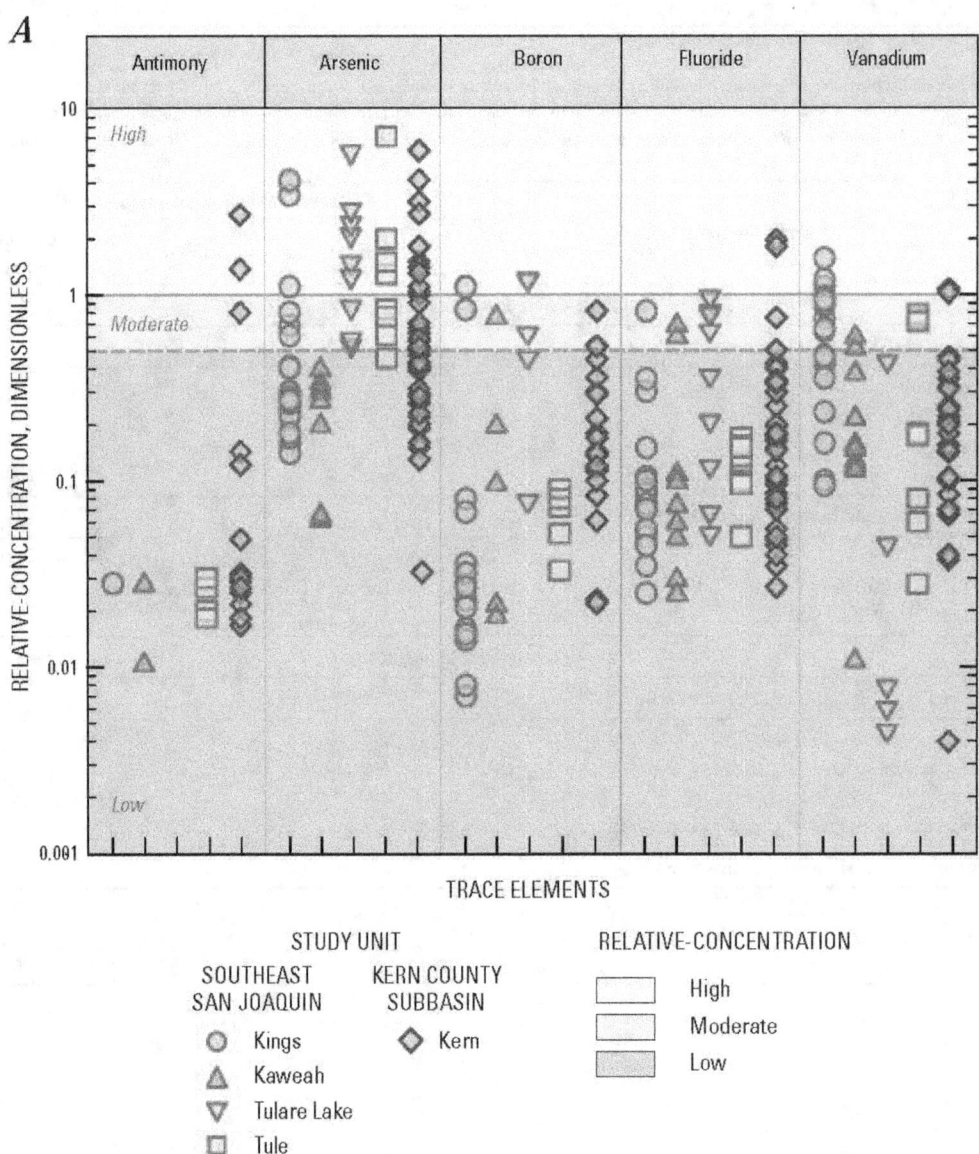

Figure 16. (A–C) Relative-concentrations of inorganic constituents with human-health-based or aesthetic benchmarks with high maximum relative-concentrations in USGS- and CDPH-grid wells in the two southern San Joaquin Valley study units, California GAMA Priority Basin Project.

B

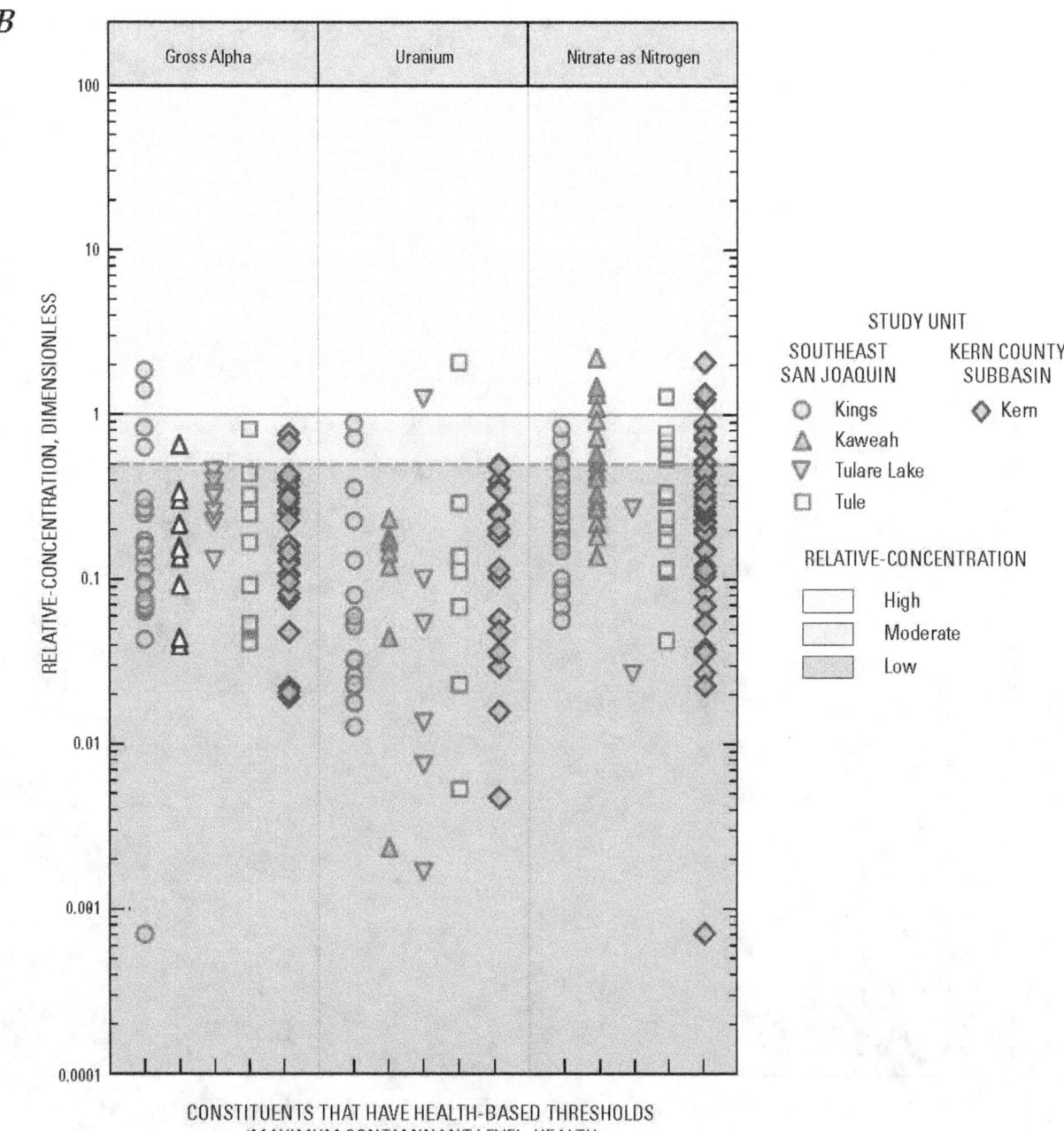

CONSTITUENTS THAT HAVE HEALTH-BASED THRESHOLDS
(MAXIMUM CONTAMINANT LEVEL, HEALTH
ADVISORY LEVEL, NOTIFICATION LEVEL)

Figure 16.—Continued

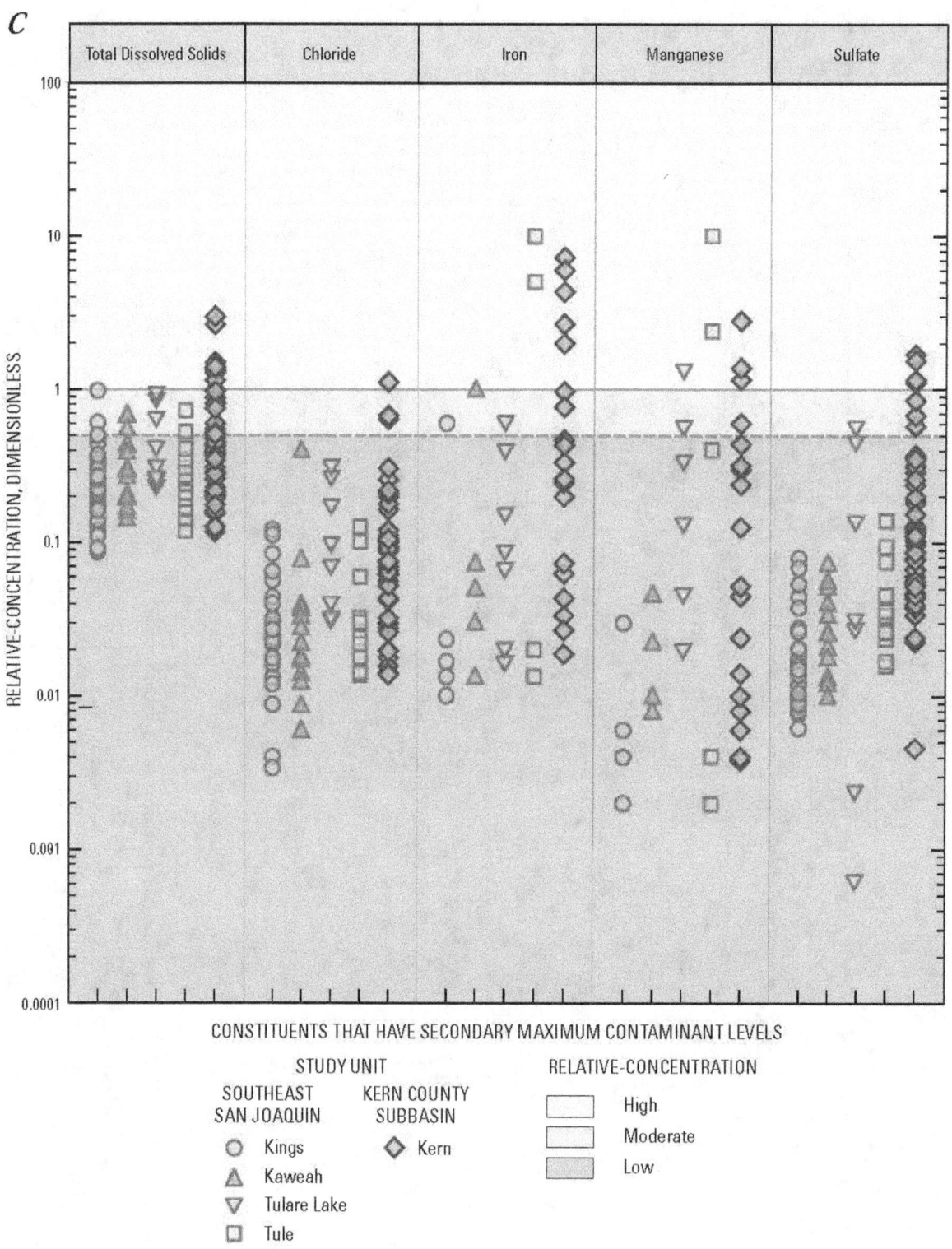

Figure 16.—Continued

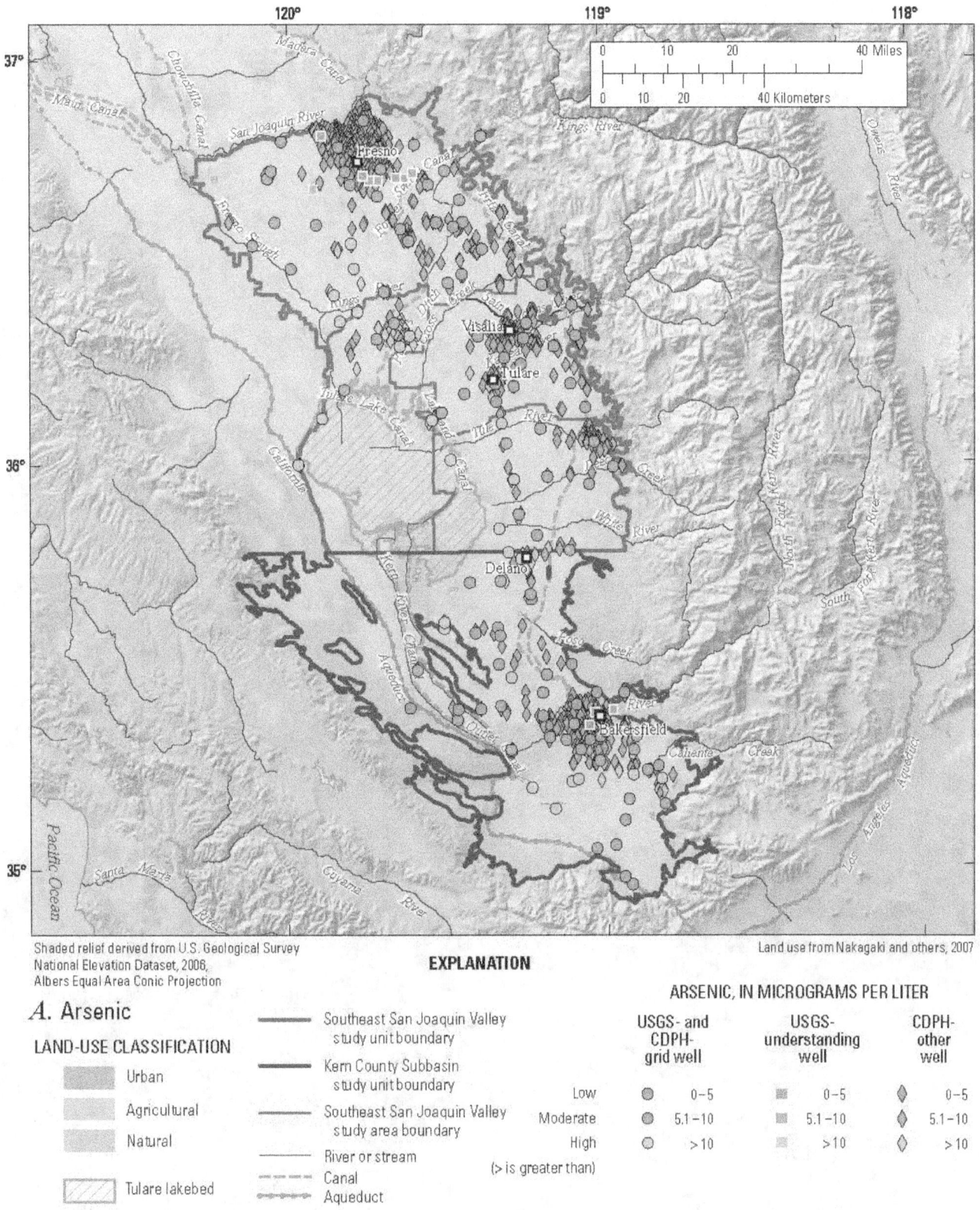

Shaded relief derived from U.S. Geological Survey
National Elevation Dataset, 2006,
Albers Equal Area Conic Projection

EXPLANATION

Land use from Nakagaki and others, 2007

A. Arsenic

LAND-USE CLASSIFICATION

Urban

Agricultural

Natural

Tulare lakebed

—— Southeast San Joaquin Valley
study unit boundary

—— Kern County Subbasin
study unit boundary

—— Southeast San Joaquin Valley
study area boundary

—— River or stream

—— Canal

—— Aqueduct

(> is greater than)

ARSENIC, IN MICROGRAMS PER LITER

	USGS- and CDPH-grid well	USGS-understanding well	CDPH-other well
Low	0–5	0–5	0–5
Moderate	5.1–10	5.1–10	5.1–10
High	>10	>10	>10

Figure 17.   *(A–K)* Relative-concentrations of selected inorganic constituents with human-health-based and aesthetic benchmarks in USGS-grid, CDPH-grid, USGS-understanding wells, and CDPH-other wells in the two southern San Joaquin Valley study units, California GAMA Priority Basin Project.

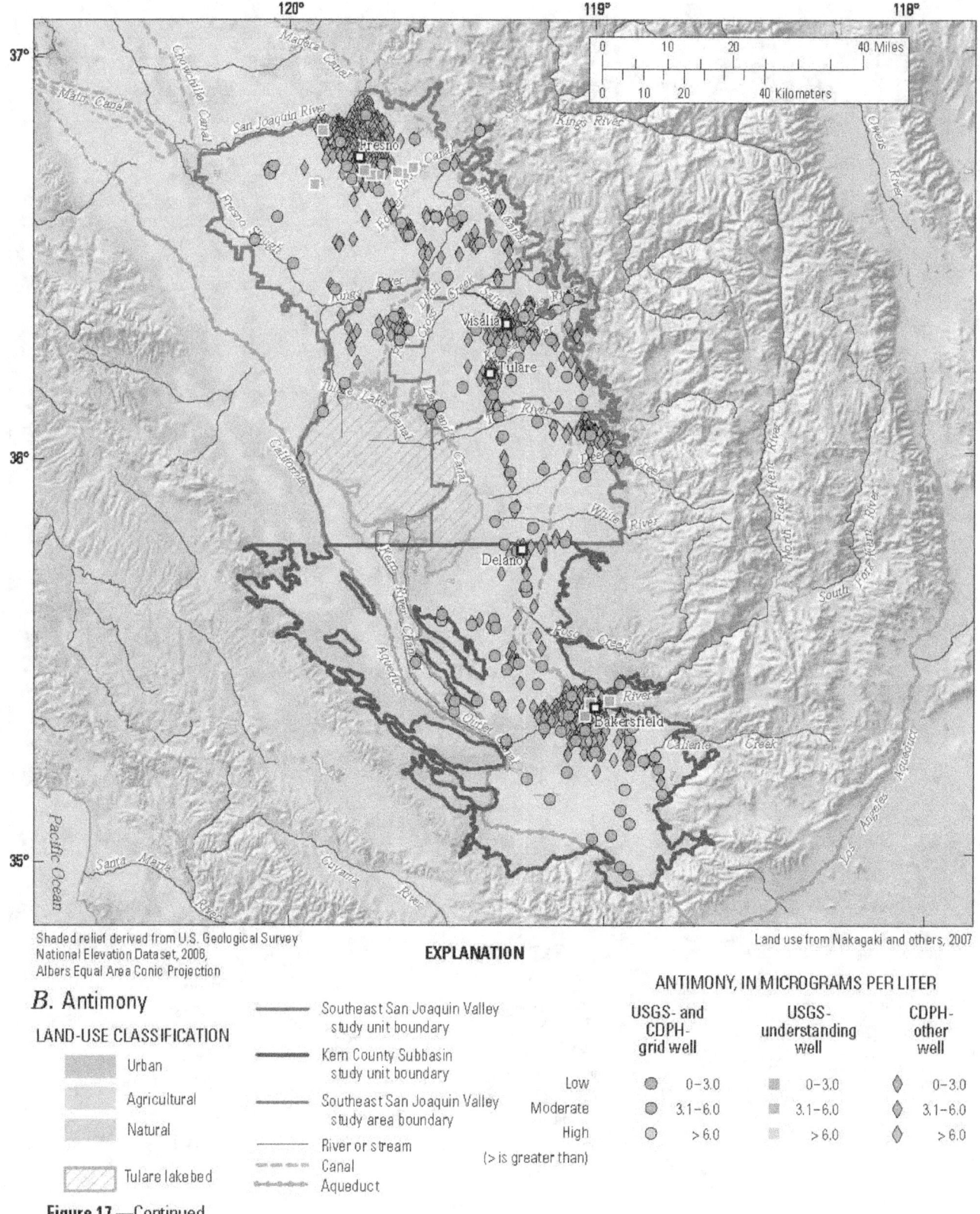

Shaded relief derived from U.S. Geological Survey
National Elevation Dataset, 2006,
Albers Equal Area Conic Projection

EXPLANATION

Land use from Nakagaki and others, 2007

B. Antimony

LAND-USE CLASSIFICATION

- Urban
- Agricultural
- Natural
- Tulare lake bed

—— Southeast San Joaquin Valley
       study unit boundary

—— Kern County Subbasin
       study unit boundary

—— Southeast San Joaquin Valley
       study area boundary

—— River or stream

---- Canal

Aqueduct

(> is greater than)

ANTIMONY, IN MICROGRAMS PER LITER

	USGS- and CDPH- grid well	USGS- understanding well	CDPH- other well
Low	0–3.0	0–3.0	0–3.0
Moderate	3.1–6.0	3.1–6.0	3.1–6.0
High	>6.0	>6.0	>6.0

Figure 17.—Continued

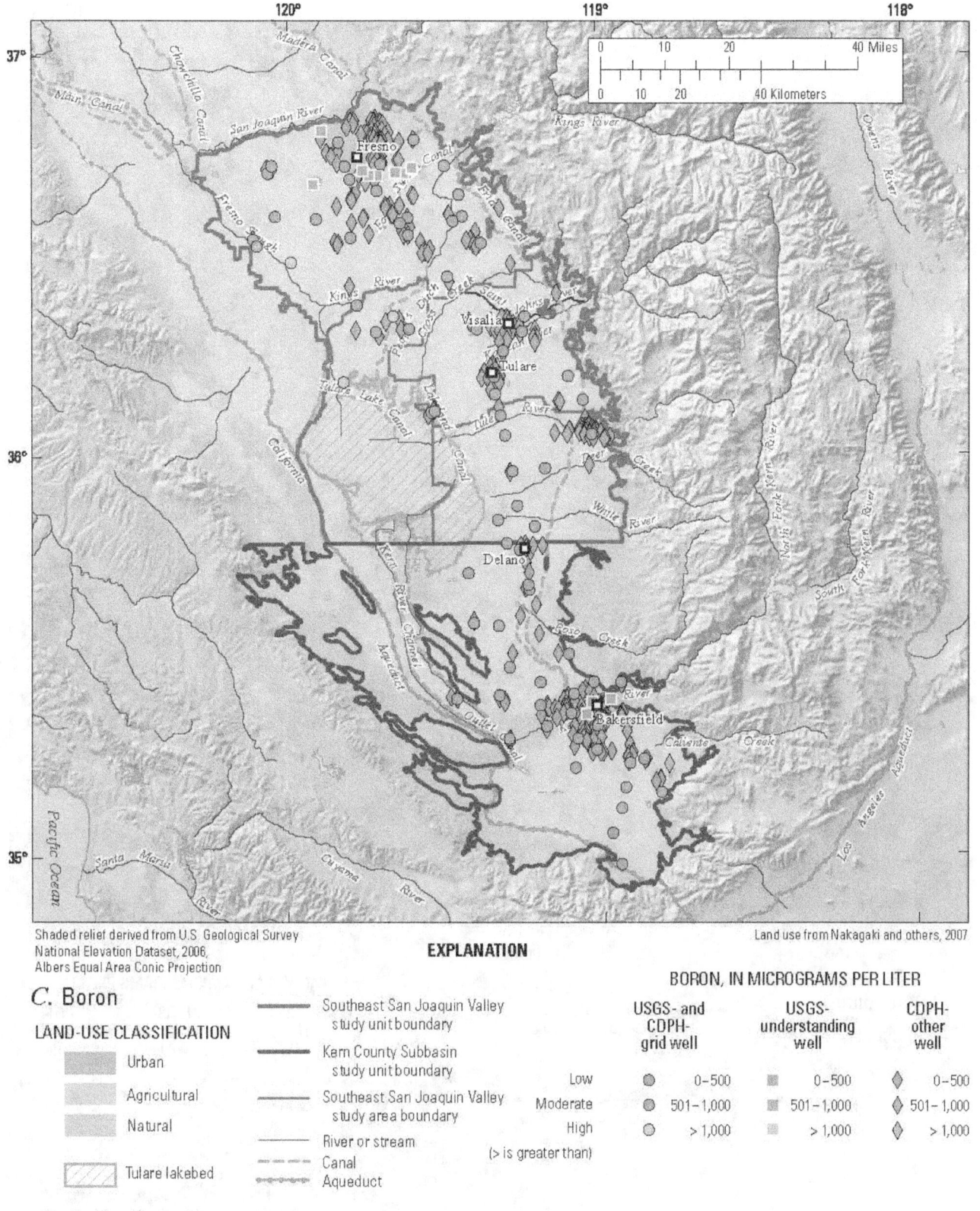

Shaded relief derived from U.S. Geological Survey
National Elevation Dataset, 2006,
Albers Equal Area Conic Projection

Land use from Nakagaki and others, 2007

EXPLANATION

C. Boron

LAND-USE CLASSIFICATION

- Urban
- Agricultural
- Natural
- Tulare lakebed

—— Southeast San Joaquin Valley
study unit boundary

━━ Kern County Subbasin
study unit boundary

—— Southeast San Joaquin Valley
study area boundary

—— River or stream

– – Canal

⟿ Aqueduct

BORON, IN MICROGRAMS PER LITER

	USGS- and CDPH-grid well	USGS-understanding well	CDPH-other well
Low	⬡ 0–500	▦ 0–500	◇ 0–500
Moderate	⬡ 501–1,000	▦ 501–1,000	◇ 501–1,000
High	⬡ >1,000	▦ >1,000	◇ >1,000

(> is greater than)

Figure 17.—Continued

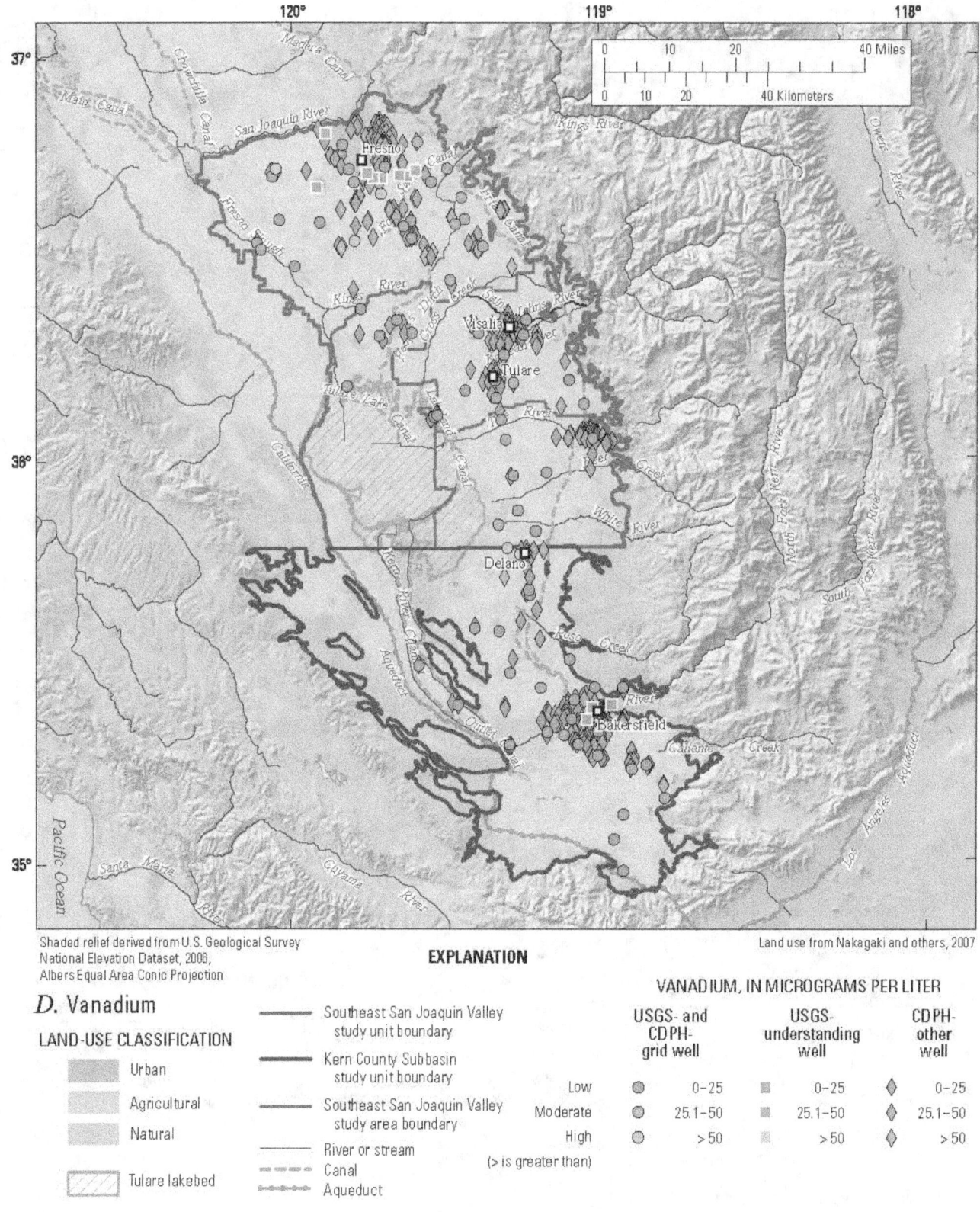

Shaded relief derived from U.S. Geological Survey
National Elevation Dataset, 2006,
Albers Equal Area Conic Projection

Land use from Nakagaki and others, 2007

EXPLANATION

D. Vanadium

LAND-USE CLASSIFICATION

- Urban
- Agricultural
- Natural
- Tulare lakebed

——— Southeast San Joaquin Valley study unit boundary

——— Kern County Subbasin study unit boundary

——— Southeast San Joaquin Valley study area boundary

——— River or stream

—— Canal

—— Aqueduct

VANADIUM, IN MICROGRAMS PER LITER

	USGS- and CDPH-grid well	USGS-understanding well	CDPH-other well
Low	0–25	0–25	0–25
Moderate	25.1–50	25.1–50	25.1–50
High	>50	>50	>50

(> is greater than)

Figure 17.—Continued

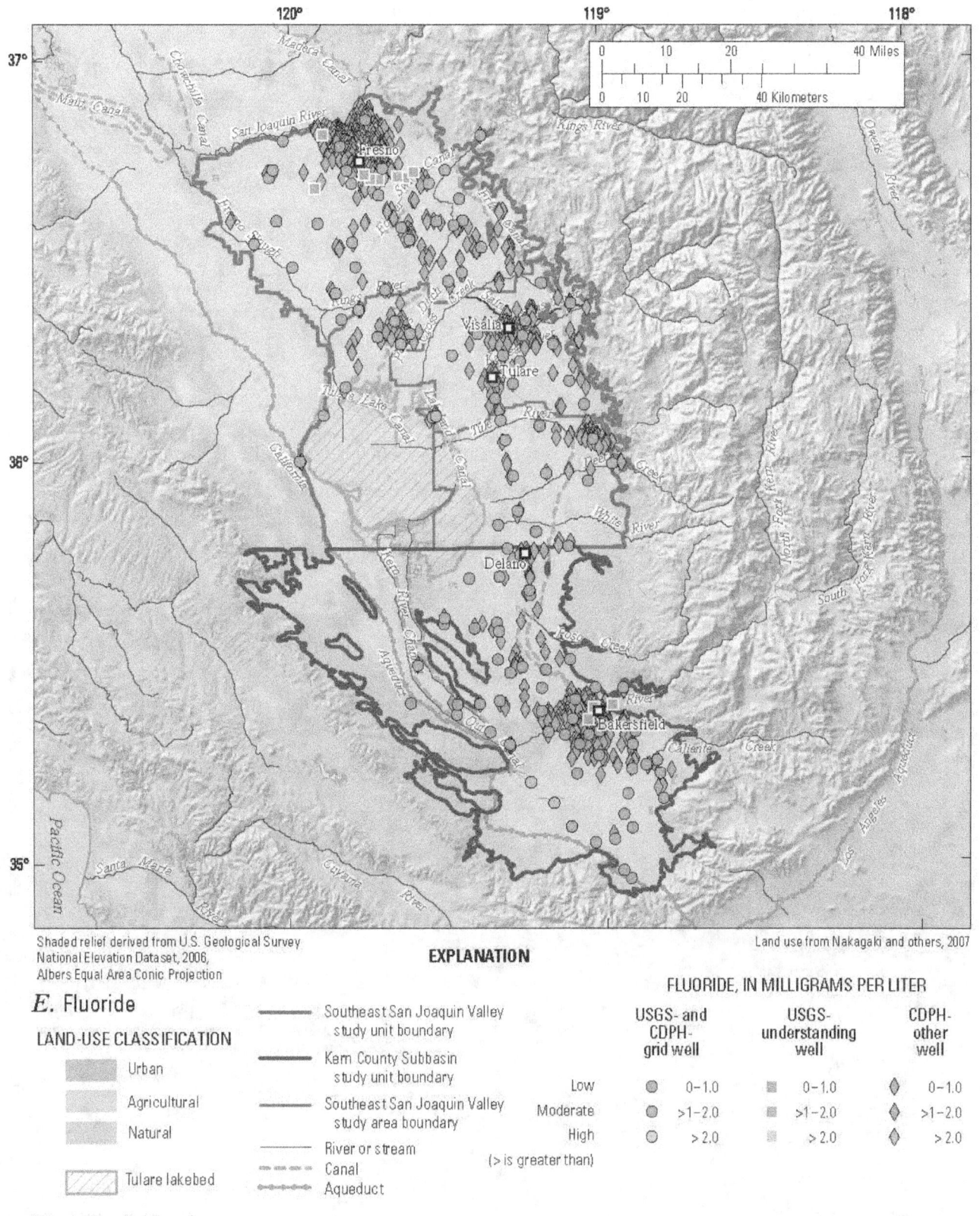

Shaded relief derived from U.S. Geological Survey
National Elevation Dataset, 2006,
Albers Equal Area Conic Projection

Land use from Nakagaki and others, 2007

EXPLANATION

E. Fluoride

LAND-USE CLASSIFICATION

Urban

Agricultural

Natural

Tulare lakebed

——— Southeast San Joaquin Valley
study unit boundary

——— Kern County Subbasin
study unit boundary

——— Southeast San Joaquin Valley
study area boundary

——— River or stream

- - - - Canal

◦-◦-◦-◦ Aqueduct

FLUORIDE, IN MILLIGRAMS PER LITER

	USGS- and CDPH- grid well	USGS- understanding well	CDPH- other well
Low	0–1.0	0–1.0	0–1.0
Moderate	>1–2.0	>1–2.0	>1–2.0
High	> 2.0	> 2.0	> 2.0

(> is greater than)

Figure 17.—Continued

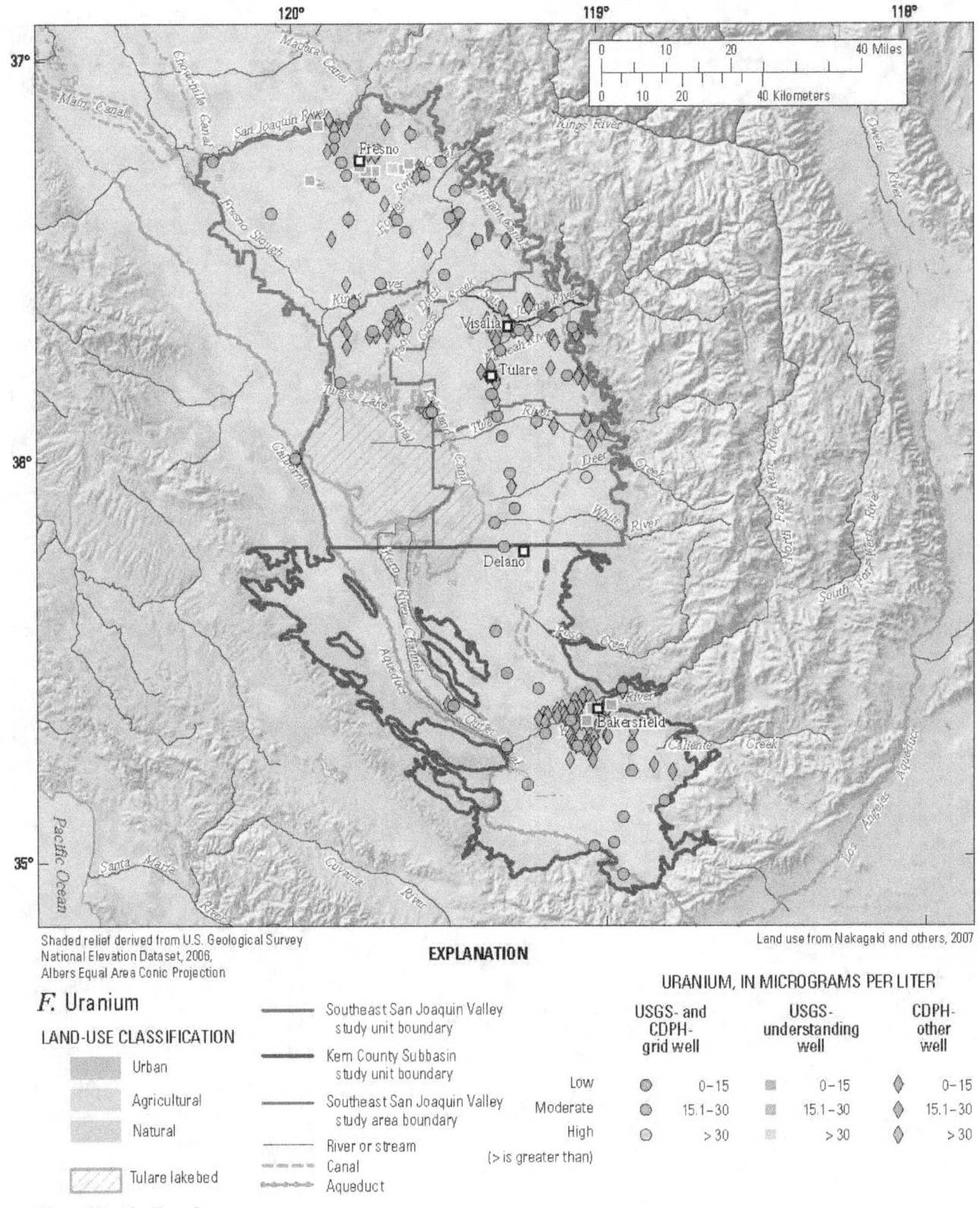

Shaded relief derived from U.S. Geological Survey
National Elevation Dataset, 2006,
Albers Equal Area Conic Projection

EXPLANATION

Land use from Nakagaki and others, 2007

F. Uranium

LAND-USE CLASSIFICATION

- Urban
- Agricultural
- Natural
- Tulare lake bed

——— Southeast San Joaquin Valley
     study unit boundary

——— Kern County Subbasin
     study unit boundary

——— Southeast San Joaquin Valley
     study area boundary

········· River or stream

——— Canal

◆◆◆◆ Aqueduct

URANIUM, IN MICROGRAMS PER LITER

	USGS- and CDPH- grid well	USGS- understanding well	CDPH- other well
Low	0–15	0–15	0–15
Moderate	15.1–30	15.1–30	15.1–30
High	>30	>30	>30

(> is greater than)

Figure 17.—Continued

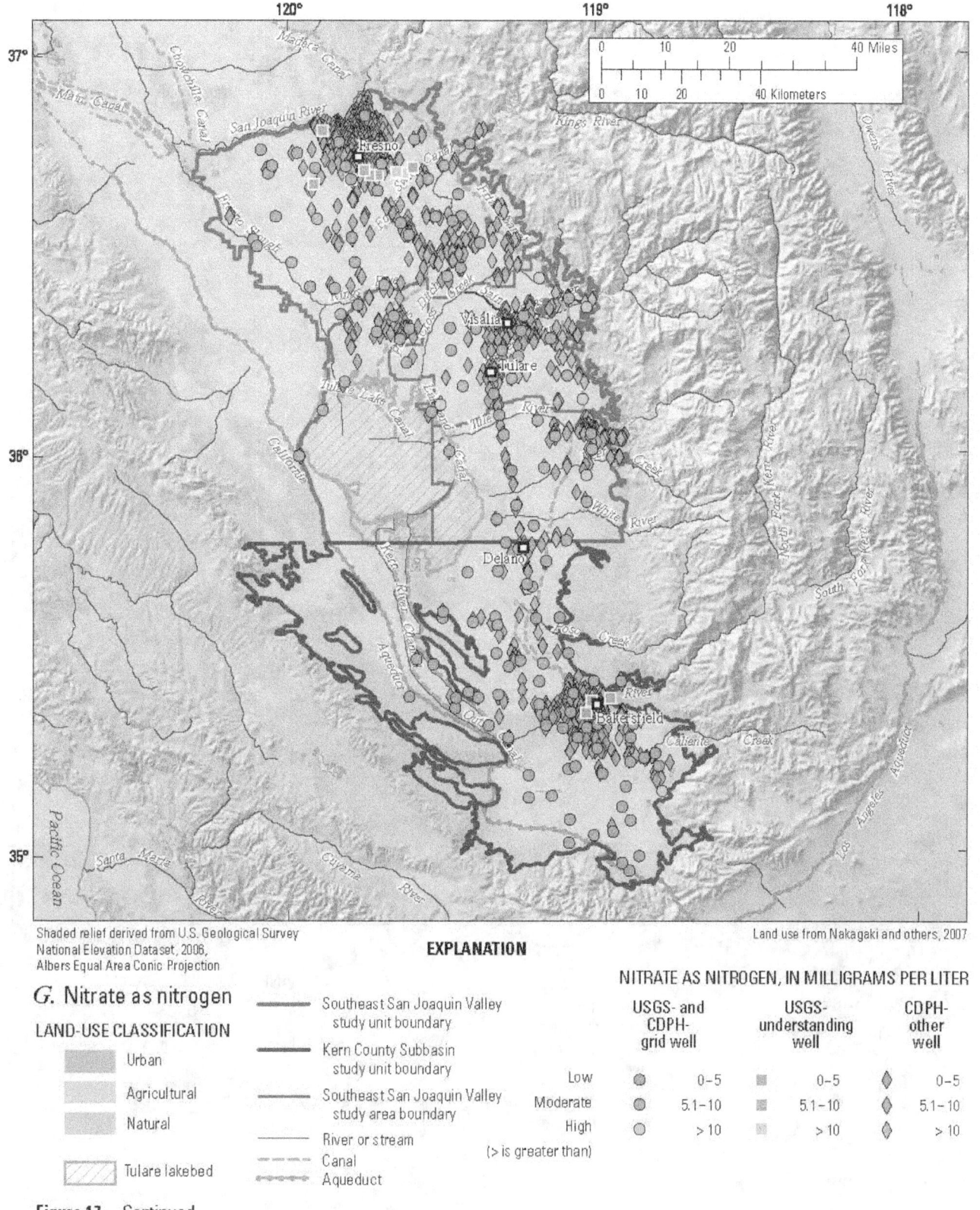

Shaded relief derived from U.S. Geological Survey
National Elevation Dataset, 2006,
Albers Equal Area Conic Projection

Land use from Nakagaki and others, 2007

EXPLANATION

G. Nitrate as nitrogen

LAND-USE CLASSIFICATION

- Urban
- Agricultural
- Natural
- Tulare lakebed

Southeast San Joaquin Valley
study unit boundary

Kern County Subbasin
study unit boundary

Southeast San Joaquin Valley
study area boundary

River or stream

Canal

Aqueduct

NITRATE AS NITROGEN, IN MILLIGRAMS PER LITER

	USGS- and CDPH- grid well	USGS- understanding well	CDPH- other well
Low	0–5	0–5	0–5
Moderate	5.1–10	5.1–10	5.1–10
High	> 10	> 10	> 10

(> is greater than)

Figure 17.—Continued

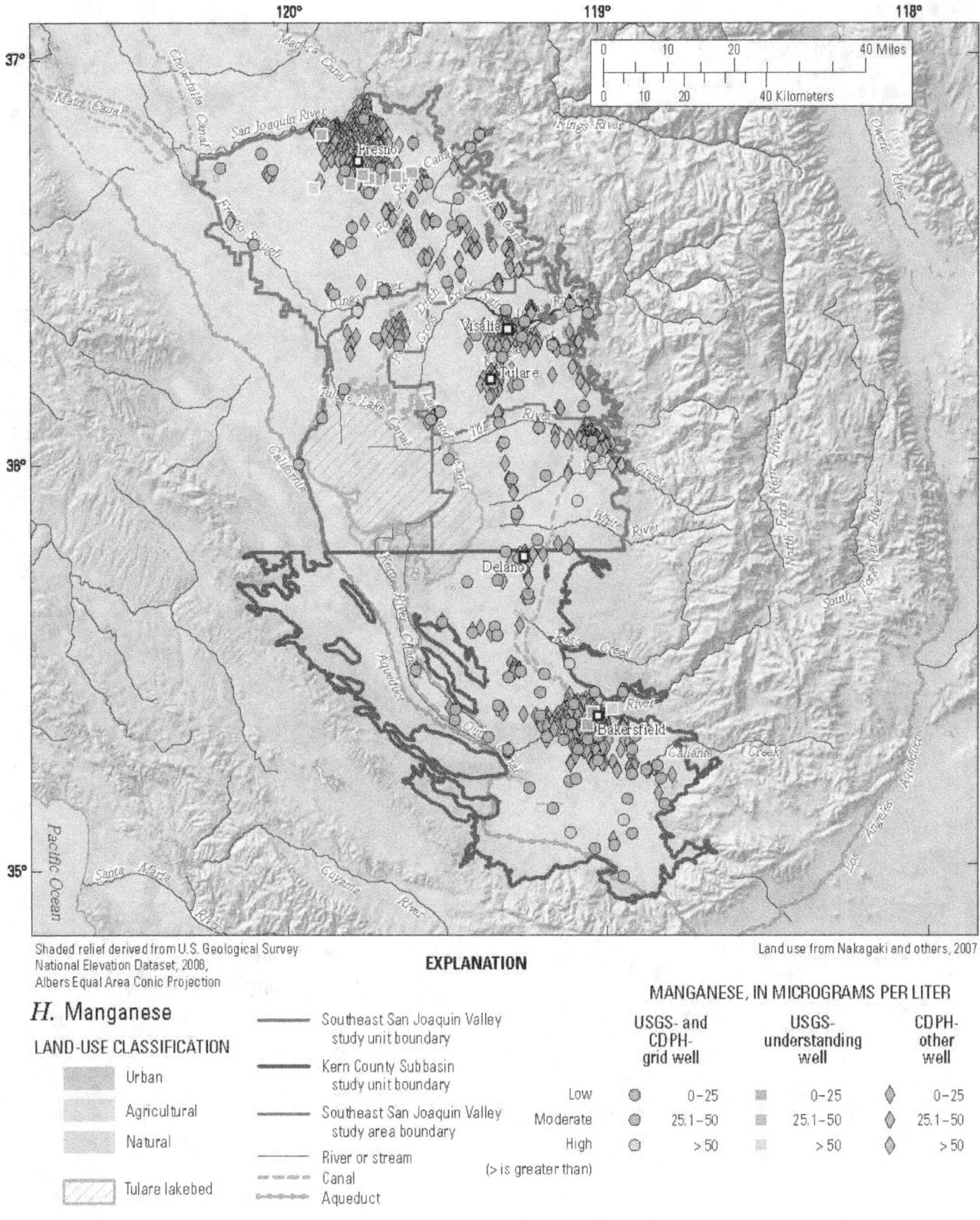

Shaded relief derived from U.S. Geological Survey
National Elevation Dataset, 2006,
Albers Equal Area Conic Projection

Land use from Nakagaki and others, 2007

EXPLANATION

H. Manganese

LAND-USE CLASSIFICATION

- Urban
- Agricultural
- Natural
- Tulare lakebed

———— Southeast San Joaquin Valley
study unit boundary

———— Kern County Subbasin
study unit boundary

———— Southeast San Joaquin Valley
study area boundary

———— River or stream

———— Canal

———— Aqueduct

(> is greater than)

MANGANESE, IN MICROGRAMS PER LITER

	USGS- and CDPH- grid well	USGS- understanding well	CDPH- other well
Low	0–25	0–25	0–25
Moderate	25.1–50	25.1–50	25.1–50
High	>50	>50	>50

Figure 17.—Continued

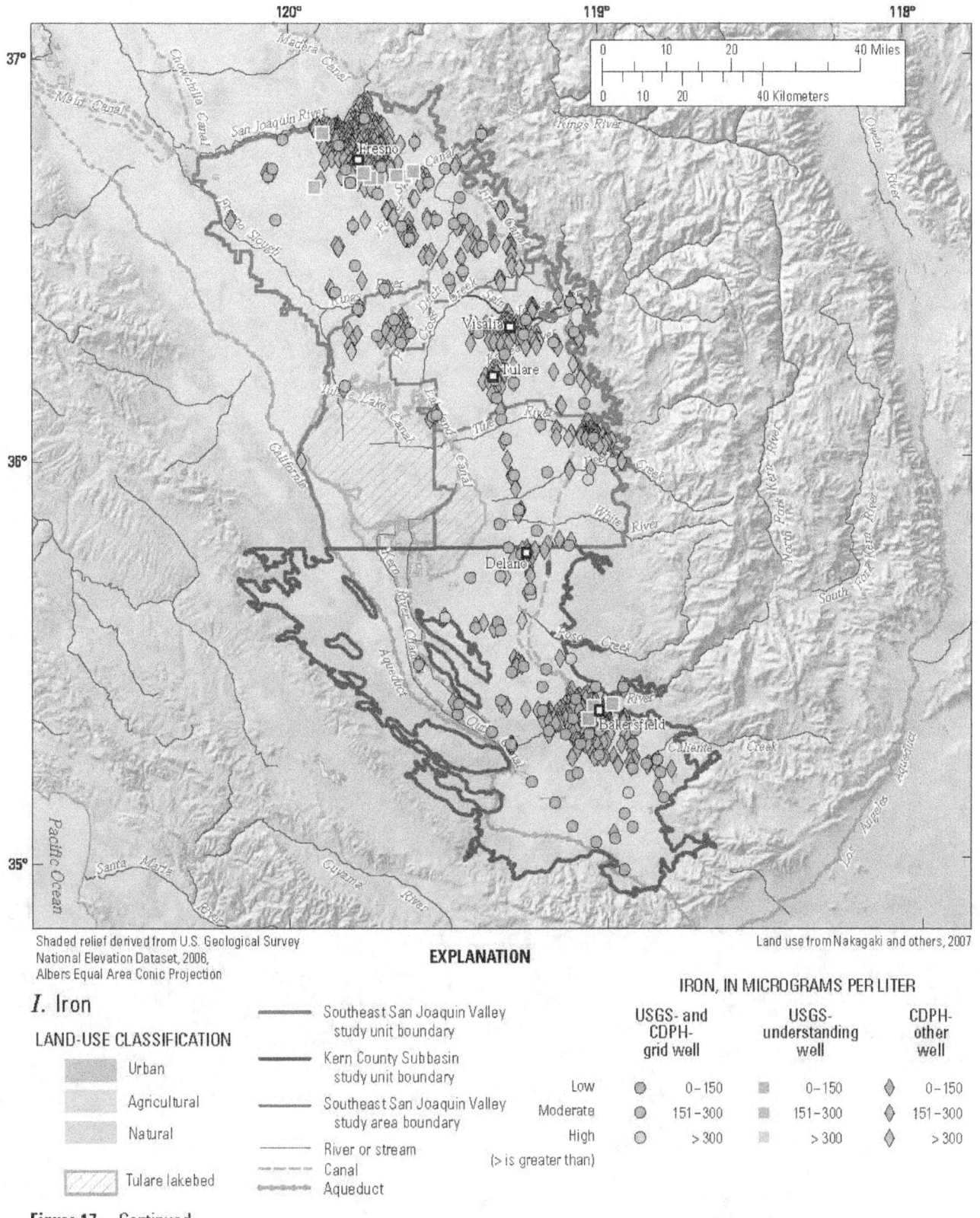

Shaded relief derived from U.S. Geological Survey
National Elevation Dataset, 2006,
Albers Equal Area Conic Projection

Land use from Nakagaki and others, 2007

EXPLANATION

I. Iron

LAND-USE CLASSIFICATION

- Urban
- Agricultural
- Natural
- Tulare lakebed

——— Southeast San Joaquin Valley
study unit boundary

——— Kern County Subbasin
study unit boundary

——— Southeast San Joaquin Valley
study area boundary

——— River or stream

—·—· Canal

—◇—◇— Aqueduct

IRON, IN MICROGRAMS PER LITER

	USGS- and CDPH- grid well	USGS- understanding well	CDPH- other well
Low	0–150	0–150	0–150
Moderate	151–300	151–300	151–300
High	>300	>300	>300

(> is greater than)

Figure 17.—Continued

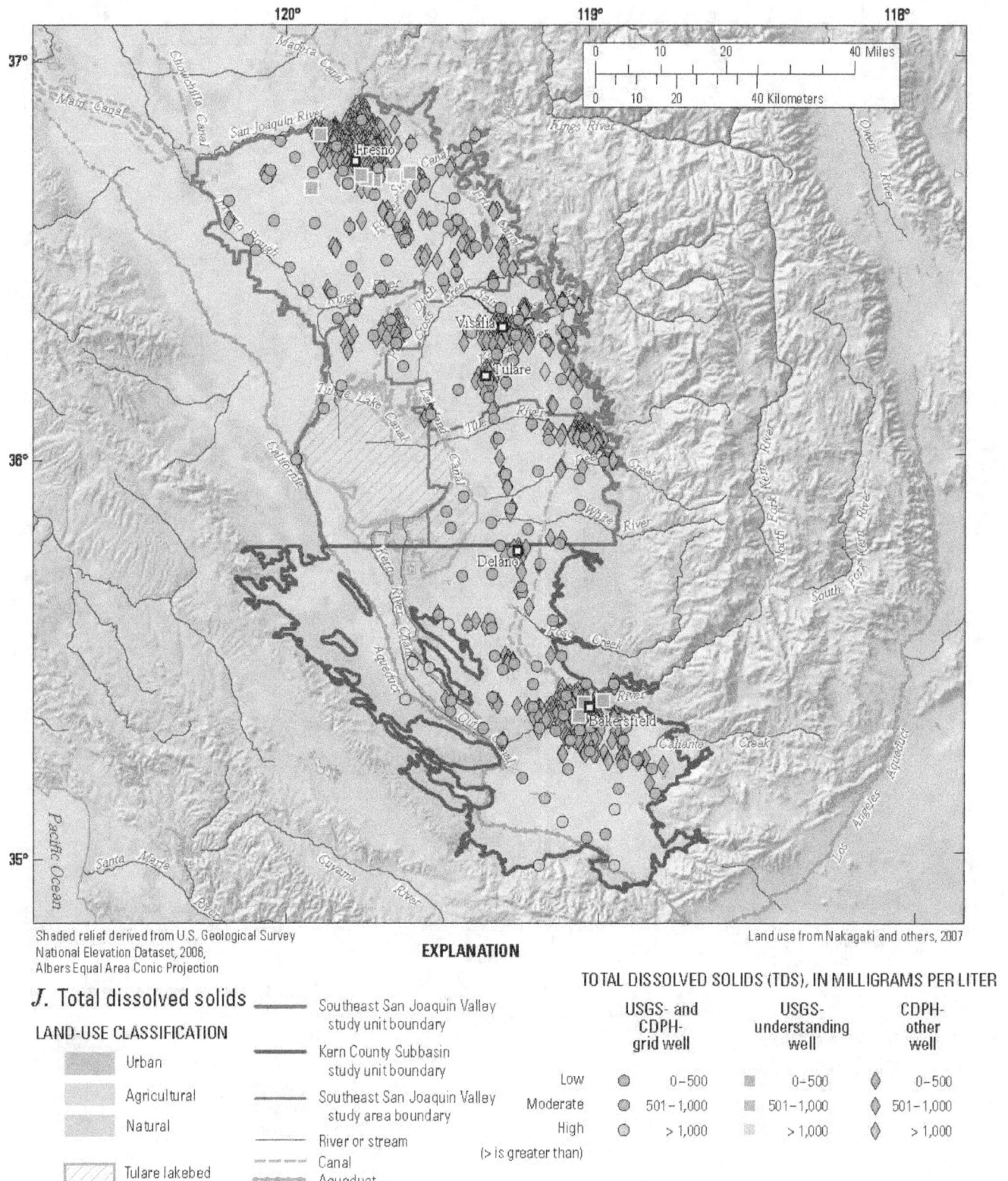

Shaded relief derived from U.S. Geological Survey
National Elevation Dataset, 2006,
Albers Equal Area Conic Projection

Land use from Nakagaki and others, 2007

EXPLANATION

J. Total dissolved solids

LAND-USE CLASSIFICATION

- Urban
- Agricultural
- Natural
- Tulare lakebed

——— Southeast San Joaquin Valley
study unit boundary

——— Kern County Subbasin
study unit boundary

——— Southeast San Joaquin Valley
study area boundary

——— River or stream

— — — Canal

—●—●— Aqueduct

TOTAL DISSOLVED SOLIDS (TDS), IN MILLIGRAMS PER LITER

	USGS- and CDPH-grid well	USGS-understanding well	CDPH-other well
Low	0–500	0–500	0–500
Moderate	501–1,000	501–1,000	501–1,000
High	>1,000	>1,000	>1,000

(> is greater than)

Figure 17.—Continued

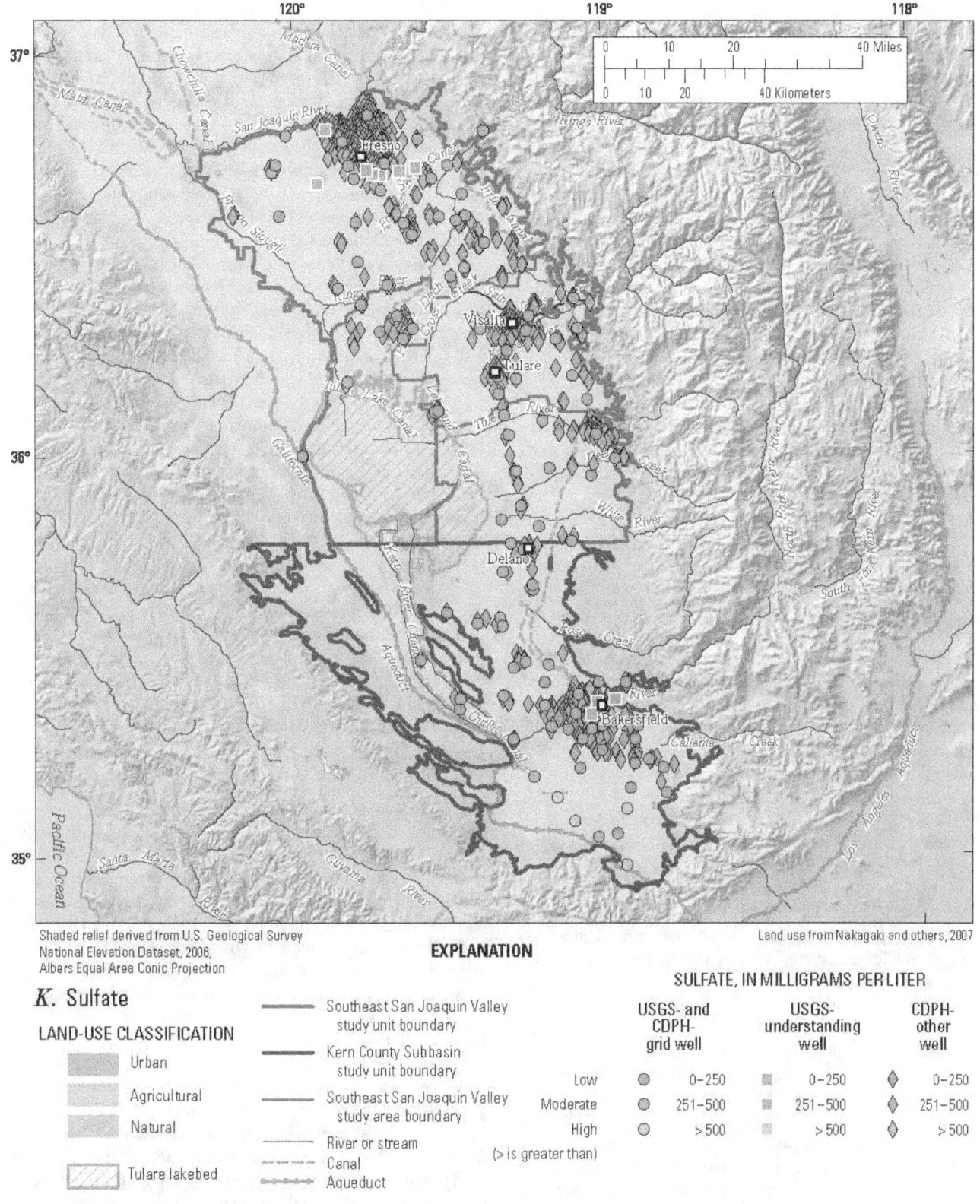

Shaded relief derived from U.S. Geological Survey
National Elevation Dataset, 2006,
Albers Equal Area Conic Projection

EXPLANATION

Land use from Nakagaki and others, 2007

K. Sulfate

LAND-USE CLASSIFICATION

- Urban
- Agricultural
- Natural
- Tulare lakebed

Southeast San Joaquin Valley
study unit boundary

Kern County Subbasin
study unit boundary

Southeast San Joaquin Valley
study area boundary

River or stream

Canal

Aqueduct

(> is greater than)

SULFATE, IN MILLIGRAMS PER LITER

	USGS- and CDPH- grid well	USGS- understanding well	CDPH- other well
Low	0–250	0–250	0–250
Moderate	251–500	251–500	251–500
High	>500	>500	>500

Figure 17.—Continued

Arsenic concentrations were significantly higher in older and deeper groundwater in the SESJ study unit (tables 6 and 12). However, arsenic concentrations were not higher in the older and deeper groundwater in the KERN study unit. Arsenic concentrations in the SESJ study unit were positively correlated with the depth to the top of the perforated interval (table 11). When the data from the Domestic Well Project is combined with the Priority Basin Project data, arsenic concentrations are higher in deep wells than in shallow wells (table 11, fig. 18A). Almost all detections with moderate or high relative-concentrations were in wells deeper than 250 ft (fig. 18A). This correlation also was detected in other Priority Basin Project study units in the San Joaquin Valley (Bennett and others, 2010; Landon and others, 2010a).

Arsenic concentrations were negatively correlated with normalized lateral position in the study units (table 11); this correlation shows that most of the high concentrations of arsenic were located in the distal part of the study units (fig. 17A). The correlation of arsenic with lateral position also may indicate that lateral position is correlated with well depth and(or) depth to top-of-perforations in the study units (tables 6A and 6B).

Arsenic also was negatively correlated with the number of septic systems located within the 500-m buffer around the well and positively correlated with percentage of agricultural land use in the SESJ study unit (table 11). The correlation with septic systems likely is a result of the correlation of septic systems to normalized lateral position and depth to top-of-perforations. Arsenic also was positively correlated with the percentage of agricultural land use in the KERN study unit (table 11). This result probably indicates that most high arsenic concentrations occur in the distal part of the study units where land use predominantly is agricultural (fig. 4) rather than agricultural land use being the cause of high arsenic concentrations.

Previous investigations of arsenic in the San Joaquin Valley (Belitz and others, 2003; Welch and others, 2006; Izbicki and others, 2008) and literature reviews (Welch and other, 2000; Stollenwerk, 2003) have indicated two mechanisms for elevated arsenic. One is the release of arsenic resulting from reductive dissolution of iron or manganese oxyhydroxides under iron- or manganese-reducing conditions. Dissolved arsenic also can increase from pH-dependent desorption of arsenic from aquifer sediments under oxic conditions; this tends to occur in groundwater with pH above 7.5 (Stollenwerk, 2003).

Evidence for pH-dependent desorption of arsenic from aquifer sediments in the SESJ study unit is the positive correlation of arsenic concentrations with pH (table 11). Almost all arsenic detections with moderate or high relative-concentrations were in samples with pH values greater than 7.6 (fig. 18B). This correlation still exists even when the shallow domestic wells are included. Arsenic may not be correlated with pH in the KERN study unit because there were very few wells with pH less than 7.6 (fig. 18B). Mobilization

of arsenic from sediments by reductive dissolution also may occur; evidence for this mechanism is supported by the presence of high concentrations of arsenic where DO concentrations were low (table 11). However, manganese- and iron-reducing conditions were not commonly found in the study units (tables D7, D8, and D9), so the correlation of arsenic with DO may indicate relations of both of these constituents to depth or pH. These observations suggest that mobilization of arsenic as a result of reducing conditions may not be as widespread as pH-dependent desorption in the southern San Joaquin Valley.

Groundwater samples in the SESJ study unit with groundwater age classified as modern contained significantly lower arsenic concentrations than groundwater samples classified as pre-modern. Other studies in the San Joaquin Valley also determined that high arsenic concentrations were associated with old groundwater (Bennett and others, 2010; Izbicki and others, 2008; Landon and others, 2010a). Groundwater age is correlated with pH in the SESJ study unit (table 5) which further supports the idea that pH-dependent desorption may be the major mechanism for the mobilization of arsenic. Arsenic may not be correlated with groundwater age in the KERN study unit because the pH for almost all of the wells in the KERN study unit was greater than 7.5 regardless of groundwater age (fig. 18A and 18B).

The correlations of arsenic concentrations with DO, pH, and well depth would explain why the highest concentrations of arsenic are in the Tulare Lake study area where wells generally are deeper, pH generally is higher, and DO generally is lower (table 5) than in the other study areas. This suggests that arsenic in SESJ and KERN primarily are derived from natural sources and not anthropogenic sources.

Antimony

Antimony is a naturally occurring semi-metallic trace element that easily combines with other elements, particularly sulfur, to form a variety of minerals. Antimony also is found in trace amounts in silver, copper, and lead ores. Anthropogenic uses of antimony include use as a flame retardant in clothes and toys, use in metal alloys, particularly in lead-acid batteries and other metallic products, and use for the clarification of specialty glasses (Carlin, 2006). It also is used in paints, ceramics, and fireworks, and as enamels for plastics, metal, and glass (Agency for Toxic Substances and Disease Registry, 1995).

Antimony was detected at high relative-concentrations in 3.6 percent and at moderate relative-concentrations in 1.8 percent of the primary aquifers in the KERN study unit (table 8; fig. 16A). Antimony was not detected at high or at moderate relative-concentrations in the SESJ study unit. The high relative-concentration of antimony occurred in CDPH-grid and CDPH-other wells in the area south of the city of Bakersfield (fig. 17B).

Table 11. Results of non-parametric (Spearman's method) correlation analysis between selected inorganic water-quality constituents and potential explanatory factors in the two southern San Joaquin Valley study units, California GAMA Priority Basin Project.

[Aquifer-scale proportions are from the grid-based method unless otherwise stated. Spearman's correlation statistic (ρ) for a significant correlation on the basis of significance level (p) less than threshold value (α) of 0.05. **Abbreviations**: TDS, total dissolved solids; MCL-US, USEPA maximum contaminant level; NL-CA, CDPH notification level; SMCL-CA, CDPH secondary maximum contaminant level; ns, no significant correlation; <, less than; GAMA, Groundwater Ambient Monitoring and Assessment Program; USEPA, U.S. Environmental Protection Agency; CDPH, California Department of Public Health]

Constituent	Benchmark type	High aquifer proportion (percent)	Grid and understanding wells used in analysis						Grid wells used in analysis			
			Depth to top-of-perforations	Well depth	pH	Dissolved oxygen	Number of septic tanks or cesspools[1]	Orchard and vineyard land use[1]	Agricultural land use[1] (percent)	Urban land use[1] (percent)	Natural land use[1] (percent)	Normalized lateral position from valley trough
			ρ: Spearman's correlation statistic/p: significance level									
			Southeast San Joaquin Valley study unit									
Arsenic[2]	MCL-US	19.0	0.341 0.006	ns	0.488 <0.001	-0.550 <0.001	-0.338 0.002	ns	0.311 0.009	ns	-0.256 0.032	-0.636 <0.001
Antimony	MCL-US	0.0	ns	ns	0.284 0.018	ns	-0.227 0.048	ns	ns	ns	ns	ns
Boron[2]	NL-CA	6.5	0.285 0.045	ns	0.366 0.006	-0.408 0.002	-0.496 <0.001	-0.250 0.050	ns	ns	ns	-0.530 <0.001
Vanadium[2]	NL-CA	6.1	ns	ns	ns	0.328 0.011	ns	ns	ns	ns	ns	0.331 0.020
Fluoride	MCL-US	0.0	0.286 0.022	0.370 0.002	0.509 <0.001	-0.448 <0.001	ns	ns	0.253 0.039	ns	ns	-0.404 <0.001
Uranium[2]	MCL-US	5.0	-0.620 <0.001	-0.483 <0.001	-0.361 0.007	0.319 0.024	0.270 0.044	ns	ns	ns	ns	ns
Nitrate[2]	MCL-US	6.3	-0.308 0.010	-0.244 0.033	-0.346 0.003	0.642 <0.001	0.309 0.002	0.341 0.001	ns	ns	ns	0.533 <0.001
Manganese[2]	SMCL-CA	4.9	ns	ns	ns	-0.314 0.011	ns	ns	ns	ns	ns	ns
Iron[2]	SMCL-CA	3.3	ns	ns	0.330 0.004	-0.355 0.004	ns	ns	ns	ns	ns	-0.395 0.002
TDS	SMCL-CA	[3]0.4	ns	ns	-0.300 0.007	ns	ns	ns	ns	ns	ns	ns
Sulfate	SMCL-CA	0.0	-0.463 <0.001	-0.372 0.002	ns	ns	ns	ns	ns	ns	ns	ns
Chloride	SMCL-CA	[3]0.3	ns	ns	ns	ns	ns	ns	ns	ns	ns	-0.288 0.025

Table 11. Results of non-parametric (Spearman's method) correlation analysis between selected inorganic water-quality constituents and potential explanatory factors in the two southern San Joaquin Valley study units, California GAMA Priority Basin Project.—Continued

[Aquifer-scale proportions are from the grid-based method unless otherwise stated. Spearman's correlation statistic (ρ) for a significant correlation on the basis of significance level (p) less than threshold value (α) of 0.05. **Abbreviations:** TDS, total dissolved solids; MCL-US, USEPA maximum contaminant level; NL-CA, CDPH notification level; SMCL-CA, CDPH secondary maximum contaminant level; ns, no significant correlation; <, less than; GAMA, Groundwater Ambient Monitoring and Assessment Program; USEPA, U.S. Environmental Protection Agency; CDPH, California Department of Public Health]

ρ: Spearman's correlation statistic/p: significance level

Constituent	Benchmark type	High aquifer proportion (percent)	Grid and understanding wells used in analysis						Grid wells used in analysis			
			Depth to top-of-perforations	Well depth	pH	Dissolved oxygen	Number of septic tanks or cesspools[1]	Orchard and vineyard land use[1]	Agricultural land use[1] (percent)	Urban land use[1] (percent)	Natural land use[1] (percent)	Normalized lateral position from valley trough
Kern County Subbasin study unit												
Arsenic[2]	MCL-US	20.0	ns	ns	ns	ns	ns	ns	0.266 0.050	ns	-0.282 0.037	-0.423 0.001
Antimony[2]	MCL-US	4.7	ns	ns	ns	0.380 0.046	ns	ns	ns	-0.322 0.023	ns	ns
Boron[2]	NL-CA	[3]2.1	ns	ns	-0.535 0.001	ns	-0.431 0.010	ns	ns	ns	ns	ns
Vanadium[2]	NL-CA	3.4	ns	ns	ns	ns	ns	0.369 0.037	ns	ns	ns	ns
Fluoride[2]	MCL-US	3.5	0.377 0.033	0.380 0.017	ns	-0.607 <0.001	-0.450 <0.001	ns	0.275 0.039	ns	ns	ns
Uranium[2]	MCL-US	[3]6.1	ns	ns	-0.426 0.042	ns	-0.464 0.026	ns	ns	ns	ns	ns
Nitrate[2]	MCL-US	4.5	ns	0.361 0.014	ns	0.482 0.003	0.297 0.013	0.289 0.015	ns	ns	ns	0.268 0.028
Manganese[2]	SMCL-CA	5.7	ns	ns	ns	-0.502 0.004	-0.362 0.006	ns	ns	ns	ns	ns
Iron[2]	SMCL-CA	9.4	ns	0.332 0.042	ns	-0.387 0.029	ns	ns	ns	ns	ns	ns
TDS[2]	SMCL-CA	14	ns	ns	-0.299 0.031	ns	-0.336 0.008	ns	0.334 0.010	ns	ns	ns
Sulfate[2]	SMCL-CA	8.3	ns	ns	ns	ns	ns	ns	0.476 0.001	ns	-0.301 0.032	ns
Chloride[2]	SMCL-CA	2.1	0.454 0.014	ns	ns	ns	ns	0.422 0.002	ns	ns	ns	ns

Table 11. Results of non-parametric (Spearman's method) correlation analysis between selected inorganic water-quality constituents and potential explanatory factors in the two southern San Joaquin Valley study units, California GAMA Priority Basin Project.—Continued

[Aquifer-scale proportions are from the grid-based method unless otherwise stated. Spearman's correlation statistic (ρ) for a significant correlation on the basis of significance level (p) less than threshold value (α) of 0.05. **Abbreviations:** TDS, total dissolved solids; MCL-US, USEPA maximum contaminant level; NL-CA, CDPH notification level; SMCL-CA, CDPH secondary maximum contaminant level; ns, no significant correlation; <, less than; GAMA, Groundwater Ambient Monitoring and Assessment Program; USEPA, U.S. Environmental Protection Agency; CDPH, California Department of Public Health]

Constituent	Benchmark type	Grid and understanding wells used in analysis							Grid wells used in analysis			
		High aquifer proportion (percent)	Depth to top-of-perforations	Well depth	pH	Dissolved oxygen	Number of septic tanks or cesspools[1]	Orchard and vineyard land use[1]	Agricultural land use[1] (percent)	Urban land use[1] (percent)	Natural land use[1] (percent)	Normalized lateral position from valley trough
								ρ: Spearman's correlation statistic/p: significance level				
Southeast San Joaquin Valley study unit with Domestic Wells included												
Arsenic	MCL-US			0.185 / 0.017	0.298 / <0.001		ns	ns				
Antimony	MCL-US			0.214 / 0.006	0.260 / <0.001		ns	-0.181 / 0.007				
Boron	NL-CA			0.334 / <0.001	0.330 / <0.001		-0.159 / 0.023	-0.196 / 0.005				
Vanadium	NL-CA			ns	ns		-0.146 / 0.036	0.187 / 0.007				
Fluoride	MCL-US			0.383 / <0.001	0.423 / <0.001		ns	-0.203 / 0.002				
Uranium	MCL-US			-0.476 / <0.001	-0.354 / 0.006		ns	ns				
Nitrate	MCL-US			-0.306 / <0.001	-0.403 / <0.001		ns	0.332 / <0.001				
Manganese	SMCL-CA			-0.256 / 0.001	-0.250 / <0.001		ns	ns				
Iron	SMCL-CA			ns	ns		ns	ns				
TDS	SMCL-CA			-0.173 / 0.022	-0.426 / <0.001		ns	0.264 / <0.001				
Sulfate	SMCL-CA			-0.293 / <0.001	-0.391 / <0.001		ns	0.330 / <0.001				
Chloride	SMCL-CA			ns	ns		-0.156 / 0.21	ns				

[1] Land-use percentages and number of septic tanks are within a circle with a radius of 500 meters around each well included in analysis.

[2] Constituents with greater than 2 percent high aquifer proportion.

[3] Based on the spatially weighted approach.

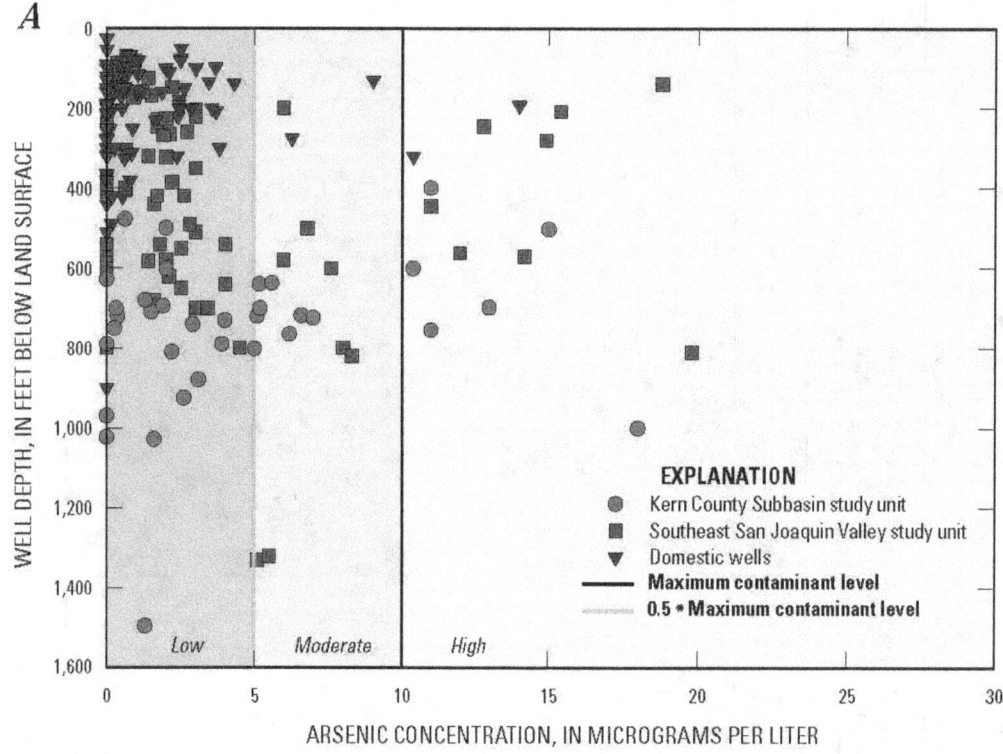

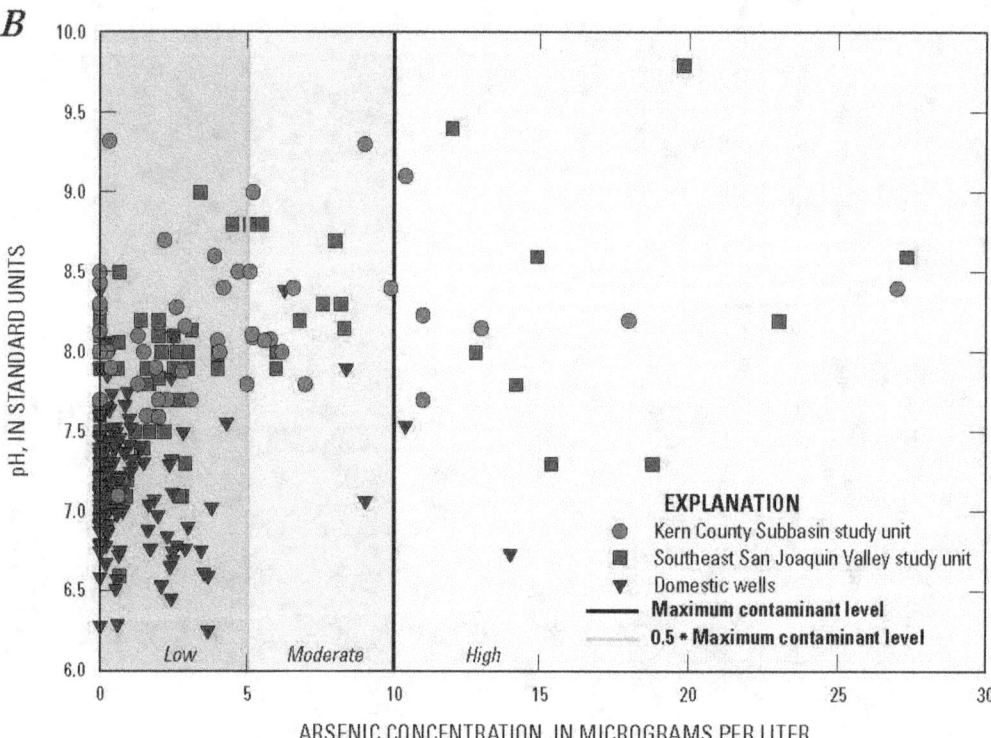

Figure 18.   Relation of arsenic concentrations to (*A*) well depth and (*B*) pH in the two southern San Joaquin Valley study units, California GAMA Priority Basin Project, and the Domestic Well Project.

Antimony concentrations, similar to arsenic, significantly increase as pH increases in the SESJ study unit. This correlation was expected as many of the chemical traits of antimony are similar to arsenic (Hem, 1985). Antimony concentrations generally increased with well depth in the SESJ study unit when data from the Domestic Well Project was included in the analysis (table 11). The correlations with depth and pH suggest that the primary source of antimony is natural. The lack of a correlation of antimony with pH or depth in the KERN study unit (table 11), where the high concentrations of antimony occurred, may be a result of pH values greater than 7.5 for almost all wells in the KERN study unit and of wells in the KERN study unit that are deeper than wells in the SESJ study unit.

Boron

Boron is a trace element that occurs in many minerals. Natural sources of boron include igneous rocks such as granite and pegmatite (as the mineral tourmaline), and evaporite minerals such as kernite and colemanite (Hem, 1970; Reimann and de Caritat, 1998). Borax, a boron-containing evaporate mineral that is mined in California, is used as a cleaning agent and therefore may be present in sewage and industrial wastes. Seawater contains 4.5 mg/L of boron (Burton, 1996).

Boron was detected at high relative-concentrations in 6.5 percent and at moderate concentrations in 6.5 percent of the primary aquifers in the SESJ study unit (table 8; fig. 16A). Boron was detected at high relative-concentrations in 2.1 percent and at moderate relative-concentrations in 9.4 percent of the primary aquifers in the KERN study unit (table 8). The few detections of boron at high and at moderate relative-concentrations in the SESJ study unit mainly occurred in the western part of the Kings study area and northern part of the Tulare Lake study area (fig. 17C). The detections of boron in the KERN study unit were in the southern part of the study unit near the Tehachapi Mountains.

Boron concentrations increase significantly with groundwater age, depth to the top-of-perforations, and increasing pH and low DO concentrations in the SESJ study unit (tables 5 and 11); in contrast, boron concentrations decrease with increasing pH in the KERN study unit (table 11). The correlations of pH and groundwater age with boron in the SESJ study unit were expected because pH and age are correlated to well depth (tables 6 and 7). The opposite relation in the KERN study unit, a negative correlation of boron with pH, may reflect that most wells in the KERN study unit are deep and do not span the range of depths and pHs as the wells in the SESJ study unit. Boron concentrations decrease significantly with the number of septic systems located within the 500-m buffer around wells in the study units (table 11). The negative correlation with septic systems suggests that boron in cleaning agents is not a major source

of boron in groundwater. Boron concentrations also decrease significantly with increasing percentage of vineyard or orchard land use in the SESJ study unit which is most likely a result of the correlation of boron with the deep wells with high pH in the western part of the SESJ study unit rather than land use as a causative relation.

Because high and moderate relative concentrations of boron were limited to the western part of the SESJ study unit and the southern part of the KERN study unit (fig. 17C), elevated concentrations of boron relative to the rest of the study units may be associated with sediments in the aquifer derived from marine deposits derived from the Coast Ranges and San Emigdio Mountains to the west and southwest of the study units (Page, 1986) and not from anthropogenic sources. These marine deposits are naturally high in boron. Saline waters, which also contain relatively high concentrations of boron (Hem, 1985) and underlie the freshwater aquifer (Page, 1986), also could potentially cause high boron concentrations by moving into the overlying continental deposits.

Vanadium

Vanadium potentially is released to groundwater from both natural and anthropogenic sources. Natural sources can be attributed to the dissolution of vanadium-rich rocks, which include mafic rocks such as basalts and gabbros (Nriagu, 1998, as cited by Wright and Belitz, 2010), and sedimentary rocks such as shale (Vine and Tourtelot, 1970, as cited by Wright and Belitz, 2010; McKelvey and others, 1986, as cited by Wright and Belitz, 2010). Vanadium has been known to be mobilized under oxic, alkaline conditions (Wright and Belitz, 2010). Anthropogenic sources of vanadium can come from waste streams associated with the ferrous metallurgy industry (International Programme on Chemical Safety, 1988, as cited by Wright and Belitz, 2010) and through the combustion of vanadium-enriched fossil fuels, primarily in the form of residual crude oil and coal (Duce and Hoffman, 1976, as cited by Wright and Belitz, 2010; Hope, 1997, as cited by Wright and Belitz, 2010). Atmospheric vanadium can be deposited to the land surface through wet and dry deposition and transported to the subsurface by infiltrating surface water.

Vanadium was detected at high relative-concentrations in 6.1 and 3.4 percent of the primary aquifers in the SESJ and KERN study units, respectively (table 8). High relative-concentrations of vanadium occurred in the Kings study area and in the KERN study unit (fig. 16A). The high and moderate relative-concentrations of vanadium primarily are located in the middle and eastern part of the SESJ study unit, especially in the Kings and Kaweah study areas (fig. 17D). The high and moderate relative-concentrations of vanadium in the KERN study unit are near the city of Delano (fig. 17D).

Vanadium concentrations increase with increasing DO and in the eastern part of the SESJ study unit (table 11). Vanadium concentrations increase with increasing percentage of vineyards or orchards in the KERN study unit (table 11). This correlation is expected as the vineyards and orchards are located on the eastern boundary of the study units at the foot of the Sierra Nevada Mountains where mafic-rich source rock can be found. The correlation with DO was expected because high concentrations of vanadium are frequently associated with oxic conditions (Wright and Belitz, 2010). Vanadium also was found to be associated with high DO in another Priority Basin Project study unit in the San Joaquin Valley north of the SESJ study unit (Landon and others, 2010a). These correlations indicate that vanadium primarily is from natural sources. Vanadium did not have a significant correlation with pH, although vanadium is frequently associated with alkaline conditions (Wright and others, 2010); this suggests that the deep aquifer sediments in the distal part of the study units where pH is high may not contain vanadium-rich rocks or that conditions are anoxic.

Fluoride

Potential sources of fluoride to groundwater are both natural and anthropogenic. Fluoride minerals can be found in igneous and sedimentary rocks. Fluoride frequently is associated with volcanic gases, and, in some areas, this may be an important source to groundwater (Hem, 1985). Fluoride often is added to drinking-water systems and toothpaste for the prevention of dental decay (Centers for Disease Control and Prevention, 2001).

Fluoride was detected at high relative-concentrations in 3.5 percent and at moderate relative-concentrations in 1.8 percent of the primary aquifers in the KERN study unit (table 8; fig. 16A). Fluoride was not detected at high relative-concentrations in the SESJ study unit but was detected at moderate relative-concentrations in 10 percent of the primary aquifers (table 8). The high relative-concentrations of fluoride in Kern were located southwest of the City of Bakersfield (fig. 17E). A cluster of detections of fluoride at moderate relative-concentrations also is located in or near the northern part of the Tulare Lake study area (fig. 17E).

Fluoride concentrations generally increase with increasing well depth, increasing depth to top-of-perforations, older groundwater age, increasing pH, and increasing percentage of agricultural land use in the study units (tables 6 and 11). The correlation of fluoride with agriculture is most likely a result of the correlation of agricultural land use with lateral position because most of the high and moderate values of fluoride occur in the distal part of the basin where the primary land use is agricultural. Fluoride concentration was higher in pre-modern than in modern-age groundwater. Groundwater age may serve as a surrogate for well depth because groundwater age and well depth are correlated (table 5). These correlations suggest high and moderate fluoride concentrations are naturally occurring and not related to human activities.

Fluoride often occurs as mineral complexes with calcium (Hem, 1985). Substitution of fluoride with hydroxide ions at mineral surfaces may occur at high pH values, and fluoride ions are more likely to adsorb to sediment surfaces at low pH (Hem, 1985). The presence of calcium complexes may limit the solubility of fluoride. Therefore, fluoride concentration often is elevated in high pH, low calcium waters. Fluoride concentrations increased with increasing pH in the SESJ study unit. Fluoride was not correlated with pH in the KERN study unit, but this may be because the pH for very few wells sampled in the KERN study unit was below 7.6 (fig. 19). Fluoride is negatively correlated with calcium in the SESJ study unit (rho = –0.324, p = 0.013) but not in the KERN study unit. The moderate relative-concentrations of fluoride in the SESJ study unit were in samples where the calcium concentration was low (less than 10 mg/L) (Burton and Belitz, 2008). The pH was greater than 7.6 and the well depths were greater than 300 ft for all the wells with moderate or high relative-concentrations of fluoride (fig. 19). The calcium concentration in wells with high relative-concentrations of fluoride was not low, but the pH was greater than 8.1, suggesting that calcium complexes are not limiting fluoride concentrations in the KERN study unit. The correlations of fluoride with lateral position and DO may be the result of the correlations of lateral position and DO with pH and well depth (table 11).

Uranium and Radioactive Constituents

Uranium-238, thorium-232, and uranium-235 are the main sources of natural radioactivity in groundwater (Hem, 1985). Uranium-238 is the most common. Gross alpha radioactivity usually consists of isotopes of radium and radon which are part of the uranium and thorium radioactive decay series (Hem, 1985).

The MCL-US (15 pCi/L) for gross alpha particle activity applies to adjusted gross alpha activity, which is equal to the measured gross alpha activity minus uranium activity (U.S. Environmental Protection Agency, 2000). Data collected by USGS-GAMA and data compiled in the CDPH database are reported as gross alpha activity without correction for uranium activity. Gross alpha is used as a screening tool to determine whether other radioactive constituents must be analyzed (California Department of Public Health, 2012). For regulatory purposes, analysis of uranium is only required if gross alpha activity is greater than 15 pCi/L; therefore, the CDPH database contains more data for gross alpha activity than for uranium. As a result, it is not always possible to calculate adjusted gross alpha activity. For this reason, gross alpha data without correction for uranium are the primary data used in the status assessments made by USGS-GAMA for Priority Basin Project study units. Examination of data from samples having USGS-GAMA data for uranium and gross alpha indicated that, in the absence of data for uranium, uncorrected gross alpha data likely provide a more accurate estimate of the aquifer-scale proportions for uranium and radioactive constituents as a class than does adjusted gross alpha (Miranda Fram, USGS California Water Science Center, written commun., 2012).

Most data for uranium in the CDPH database are reported as activities in units of picocuries per liter, and the majority of uranium data gathered by USGS-GAMA are reported as concentrations in units of micrograms per liter. The factor used to convert uranium mass concentration to uranium activity depends on the isotopic composition of the uranium (U.S. Environmental Protection Agency, 2000). This report uses a conversion factor of 0.79.

Radioactive constituents were detected at high relative-concentrations in 6.9 percent, at moderate

relative-concentrations in 6.9 percent, and at low relative-concentrations or not detected in about 86 percent of the primary aquifers in the SESJ study unit (table 10). The radioactive constituents with high relative-concentrations were uranium and unadjusted gross alpha radioactivity (table 8; fig. 16B). Radioactive constituents were detected at high relative-concentrations in 6.1 percent, moderate in 13 percent, and low or not detected in 81 percent of the primary aquifers in the KERN study unit (table 10). The radioactive constituents with high relative-concentrations in the KERN study unit were radium, uranium, and unadjusted gross alpha radioactivity (table 8).

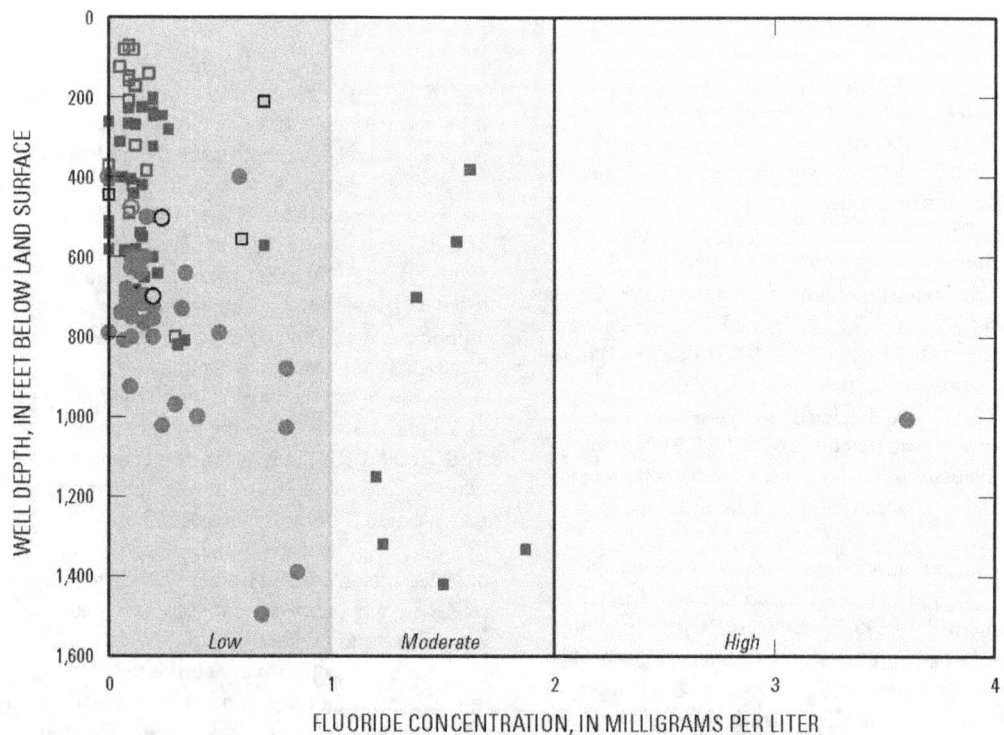

EXPLANATION

☐ Southeast San Joaquin Valley study unit, pH less than 7.6
■ Southeast San Joaquin Valley study unit, pH equal to or greater than 7.6
☐ Southeast San Joaquin Valley study unit, pH unknown
○ Kern County Subbasin study unit, pH less than 7.6
● Kern County Subbasin study unit, pH equal to or greater than 7.6
◉ Kern County Subbasin study unit, pH unknown

——— Maximum contaminant level
--- 0.5 * Maximum contaminant level

Figure 19.   Relation of fluoride concentrations to well depth and pH in the two southern San Joaquin Valley study units, California GAMA Priority Basin Project.

Uranium and Gross Alpha Radioactivity

Unadjusted gross alpha radioactivity was detected at high relative-concentrations in 3.8 percent and at moderate relative-concentrations in 7.5 percent of the primary aquifers in the SESJ study unit (table 8). The high relative-concentrations of gross alpha radioactivity were in the Kings study area (fig. 16B). Gross alpha radioactivity was detected at high relative-concentrations in 4.0 percent of the primary aquifers in the KERN study unit and at moderate relative-concentrations in 13 percent of the primary aquifers (table 8).

Gross alpha radioactivity was positively correlated with uranium (rho = 0.716, p = <0.001). This result suggests gross alpha radioactivity measurements may be considered a surrogate for uranium concentrations in the study units. For this reason, gross alpha radioactivity is not discussed further in this report.

Uranium was detected at high relative-concentrations in 5.0 percent and at moderate relative-concentrations in 5.0 percent of the primary aquifers in the SESJ study unit (table 8). The uranium detections at high and at moderate relative-concentrations were detected in grid wells in the Tulare Lake and Tule study areas (fig. 16B); additional uranium detections at moderate relative-concentrations were in the Kings study area. High relative-concentrations also were detected in CDPH-other wells and in one USGS-understanding well in the Kings study area and in a CDPH-other well in the Kaweah study area (fig. 17F).

Uranium was detected at high relative-concentrations in 6.1 percent of the primary aquifers in the KERN study unit. Uranium detections in CDPH-other wells at high or at moderate relative-concentrations are around the city of Bakersfield (fig. 17F).

Uranium concentrations were negatively correlated with pH (table 11), and positively correlated with calcium (SESJ study unit, rho=0.678, p=<0.001; KERN study unit, not significant) and alkalinity (SESJ study unit, rho=0.414, p=0.001; KERN study unit, rho=0.693, p=<0.001, fig. 20B). The results for the SESJ and KERN study units mirror the results of a local-scale investigation in an area to the north near the city of Modesto (Jurgens and others, 2008), a regional investigation in the eastern San Joaquin Valley (Jurgens and others, 2010), and another Priority Basin Project study unit (Landon and others, 2010a). Elevated uranium in shallow groundwater was attributed by Jurgens and others (2008, 2010) to the enhanced desorption of uranium from sediments by irrigation and urban recharge having high bicarbonate (alkalinity) concentrations.

Uranium concentrations were significantly greater in modern- and mixed-age groundwater than pre-modern-age groundwater (table 5), and were negatively correlated with well depths and depth to the top-of-perforations (table 11; fig. 20A) in the SESJ study unit. The association of high uranium with modern and mixed ages is consistent with the mobilization of naturally occurring uranium by irrigation and urban recharge in the shallow part of the aquifer. The lack of correlation with well depth (or depth to top-of-perforations) in the KERN study unit may be because very few wells with uranium data had a well depth less than 400 ft (fig. 20A).

Nutrients

Nutrients with human-health benchmarks, as a class, were detected at high relative-concentrations (for one or more constituents) in the study units. Nutrients were detected at high relative-concentrations in both SESJ and KERN (table 10). The only nutrient detected at high relative-concentrations in the study units was nitrate plus nitrite (hereinafter referred to as nitrate) (table 8).

Nitrate

Nitrogen in groundwater occurs in the forms of dissolved nitrate, nitrite, or ammonia. Certain bacteria and algae naturally convert nitrogen from the atmosphere to nitrate, which is an important nutrient for plants. Nitrate also is present in precipitation (Hem, 1970). Anthropogenic sources of nitrate include its application as a fertilizer for agriculture and production by livestock of nitrogenous waste that can leach to groundwater when animals are present in concentrated numbers (Hem, 1985). Septic systems also contain nitrogenous waste that may leach into groundwater.

Nitrate was detected at high relative-concentrations in 6.3 and 4.5 percent of the primary aquifers in the SESJ and KERN study units, respectively (table 8). High relative-concentrations of nitrate occurred in grid wells in the Kaweah and Tule study areas and in the KERN study unit (fig. 16B). High relative-concentrations of nitrate also were detected in some CDPH-other wells. Most high and moderate relative-concentrations of nitrate were detected in the eastern part of the study units (fig. 17G).

Nitrate was positively correlated with dissolved oxygen, lateral position, orchard and vineyard land use, and septic systems (table 11) in the study units. The correlation of nitrate to septic systems was not unexpected because septic systems can be a source of nitrate. The relation of nitrate to normalized lateral position and DO partially may indicate that nitrate is a redox-sensitive constituent that is removed from groundwater in a reducing environment, and this relation also may occur because wells become deeper toward the valley trough. Reducing conditions mostly exist toward the distal part of the southern San Joaquin Valley, at the low end of normalized lateral position values.

Nitrate concentration was negatively correlated with well depth (fig. 21A), depth to the top-of-perforations, and pH (table 11) in the SESJ study unit. High and moderate relative-concentrations of nitrate were detected in the shallow USGS-understanding wells in the Kings study area (fig. 17G).

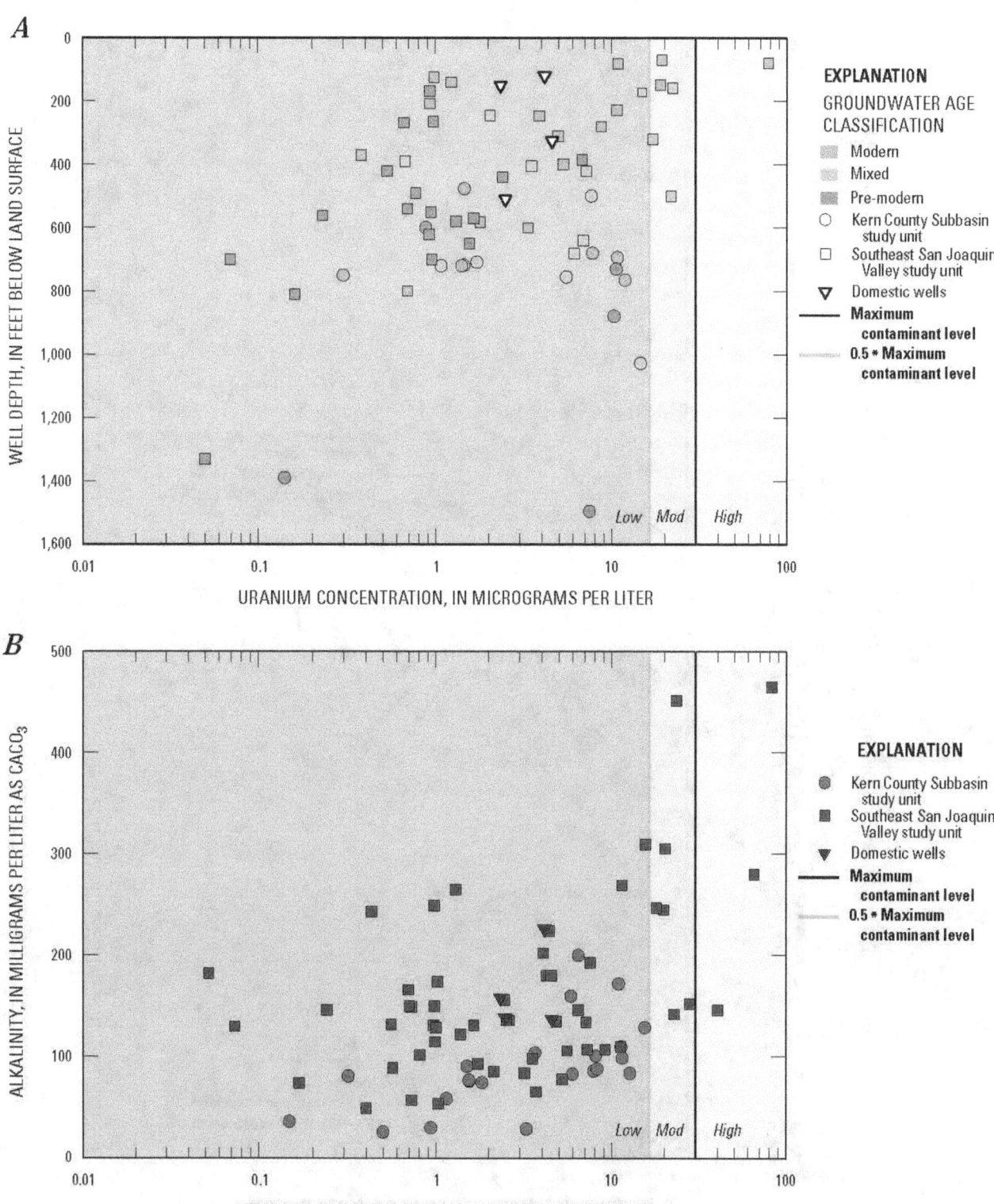

Figure 20. Relation of uranium concentrations to (A) well depth, groundwater age classification, and (B) alkalinity in the two southern San Joaquin Valley study units, California GAMA Priority Basin Project.

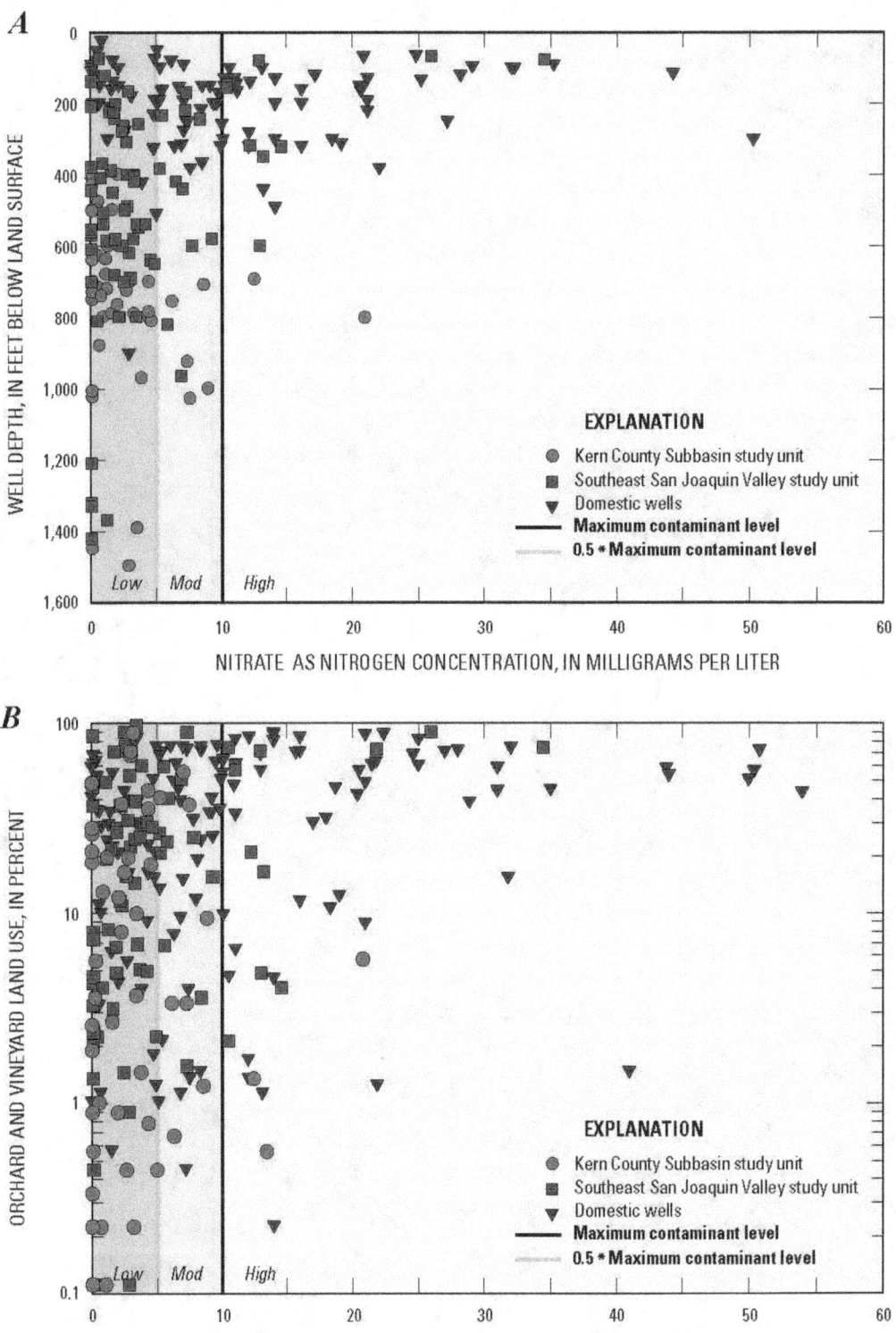

Figure 21.   Relation of nitrate to (*A*) well depth and (*B*) percentage of orchard and vineyard land use in the two southern San Joaquin Valley study units, California GAMA Priority Basin Project.

Nitrate in groundwater has been studied extensively in the eastern San Joaquin Valley (for example, Dubrovsky and others, 1998; Burow and others, 2008b). Results of these investigations of nitrate in the San Joaquin Valley have shown positive correlations between nitrate concentrations in relatively shallow parts of the aquifer and percentage of agricultural land use (Burow and others, 1998a, 1998b, 2007). This study, similar to other Priority Basin Project study units in the northern part of the San Joaquin Valley (Bennett and others, 2010; Landon and others, 2010a), did not show a direct correlation of nitrate concentrations with agricultural land use. However, in this study and that of Landon and others (2010a), nitrate is correlated with a specific type of agricultural land use, percentage of orchard and vineyard land use (fig. 21B; table 11). Percentage of orchard and vineyard land use is positively correlated with normalized lateral position and is most common in the eastern part of the study units.

Nitrate concentrations are significantly higher in shallow wells than in deep wells in the SESJ study unit (table 11). More evidence for the negative relation of nitrate with well depth is supplied by comparing the data from the Domestic Well Project and the grid wells from the study units. The domestic wells were significantly shallower than the grid wells from either study unit (fig. 7; table 5), and the concentration of nitrate is higher in domestic wells than in grid wells from the study units (table 5; fig. 21A). Nitrate concentrations in the KERN study unit generally are higher in some of the deep wells than in the SESJ study unit and in the domestic wells (table 11; fig. 21B). This result may be affected by the lack of wells less than 400 ft below land surface where most of the high concentrations of nitrate were found in other parts of the San Joaquin Valley. Nitrate concentration was not significantly correlated with groundwater age in the study units; this was unexpected because well depth and groundwater age are significantly correlated (table 5). The correlations of nitrate with well depth, septic systems, and orchard and vineyard land use suggest elevated nitrate concentrations are related to human activities.

Some of the explanatory factors related to nitrate are themselves related—DO with depth to the top-of-perforations (tables 6A,B), and lateral position and pH (table 6B). The correlations between explanatory factors could affect the correlations between nitrate and the explanatory factor. For example, the correlations of nitrate concentrations with well depth may be strengthened by the correlation of well depth with DO. Also, the relation of nitrate to lateral position may be strengthened by the correlation of nitrate with orchard and vineyard land use which is located on the eastern margin on the study units where DO is higher.

Inorganics with Aesthetic Benchmarks

As a class, inorganics with aesthetic benchmarks (SMCLs) were detected at high relative-concentrations (for one or more constituents) in the study units. Inorganics with SMCLs were detected at high relative-concentrations in 6.6 percent, at moderate relative-concentrations in 13 percent, and at low relative-concentrations or not detected in 81 percent of the primary aquifers in the SESJ study unit (table 10). The constituents with high relative-concentrations in the SESJ study unit were manganese and iron (table 8). Relative-concentrations of inorganic constituents with SMCLs were high in 22 percent, moderate in 17 percent, and low in 61 percent of the primary aquifers in the KERN study unit (table 10). The constituents with high relative-concentrations in the KERN study unit were manganese, iron, TDS, sulfate, and chloride (table 8).

For USGS- and CDPH-grid wells without a measured TDS value, TDS was calculated from specific conductance (SC) using a linear regression equation. Two linear regression equations, one for each study unit, were developed from USGS-grid and understanding wells having measured SC and TDS data. SC, an electrical measure of TDS, was available in all 130 USGS-grid and 19 USGS-understanding wells, whereas laboratory-measured TDS data (as residue on evaporation) were available for only 61 of these wells. The linear regression equation (TDS = (0.657*SC)–8.503) was developed from data for 44 USGS-grid and understanding wells in the SESJ study unit. The linear regression equation (TDS = (0.699*SC)–28.59) was developed from data for 17 USGS-grid and understanding wells in the KERN study unit. The predicted TDS using the regression equations closely matched measured TDS ($r^2 > 0.98$) for both study units.

Manganese and Iron

Potential natural sources of manganese and iron to groundwater include the dissolution of igneous and metamorphic rocks as well as various secondary minerals (Hem, 1970) which can be mobilized under reducing or low pH conditions. Potential anthropogenic sources of these constituents to groundwater include effluents associated with the steel and mining industries (Reimann and de Caritat, 1998), and soil amendments, in the form of manganese and iron sulfates, that are added to deficient soils in order to stimulate crop growth.

Manganese was detected at high relative-concentrations in 4.9 percent and 5.7 percent of the primary aquifers in the SESJ and KERN study units, respectively (table 8). High relative-concentrations of manganese occurred in USGS- and CDPH-grid wells in the Tulare Lake and Tule study areas in the SESJ and in KERN study units (fig. 16C). High relative-concentrations of manganese also were detected in some CDPH-other wells in the Kings and Kaweah study areas and one USGS-understanding well in the Kings study area. Most of the high and moderate relative-concentrations of manganese were near the eastern boundary of the SESJ and KERN study units, the southern boundary of the KERN study unit, or near the valley center in the Kings study area (fig. 17H).

Iron was detected at high relative-concentrations in 3.3 percent and 9.4 percent of the primary aquifers in the SESJ and KERN study units, respectively (table 8). High relative-concentrations of iron occurred in USGS- and CDPH-grid wells in the Tule study area and in the KERN study unit (fig. 16C). Distributions of high relative-concentrations of iron were similar to high relative-concentrations of manganese. In addition, high relative-concentrations of iron were detected in CDPH-other wells in the northern part of the Tulare Lake study area and south of the city of Delano in the KERN study area (fig. 17I).

Manganese and iron were negatively correlated to DO in the SESJ and KERN study units (table 11). Concentrations of manganese were higher in pre-modern-age groundwater than in mixed or modern-age groundwater in the SESJ study unit; concentrations of iron were higher in pre-modern-age groundwater than in modern-age groundwater in the SESJ and KERN study units (table 5). The higher concentrations of manganese and iron in pre-modern-age groundwater suggest that the primary source of manganese and iron is natural. Most of the remaining significant correlations of manganese (depth to top-of-perforations) or iron (normalized lateral position, wells depth, pH) with other explanatory factors observed in the KERN and SESJ study units (table 11) may be affected by their relations with DO (table 6A and B). The DO in deep wells in the SESJ study unit generally was low, and pH was high. Unexpectedly, manganese was negatively correlated to septic system density in the KERN study unit (table 11); however, this apparent correlation likely indicates relations of manganese and septic system density to other explanatory factors rather than a direct relation.

Total Dissolved Solids

Natural sources of TDS include (1) mixing of groundwater with deep saline groundwater that is affected by interactions with deep marine or lacustrine sediments, (2) concentration by evaporation in discharge areas, and(or) (3) rock/water interaction. Potential anthropogenic sources of TDS to groundwater include agricultural and urban irrigation, evaporation, disposal of wastewater and industrial effluent, and leaking water and sewer pipes.

TDS was detected at high relative-concentrations in 14 percent and at moderate relative-concentrations in 17 percent of the primary aquifers in the KERN study unit (table 8; fig. 16C). TDS was detected at high relative-concentrations in 0.4 percent and at moderate relative-concentrations in 13 percent of the primary aquifers in the SESJ study unit (table 8). The high and moderate relative-concentrations of TDS occurred in CDPH-grid and CDPH-other wells in the area south of the city of Bakersfield (fig. 17J), near the Tehachapi and San Emigdio Mountains, and in the distal part of the KERN study unit. Moderate relative-concentrations of TDS were detected on the eastern and western boundaries of the SESJ study unit (fig. 17J).

In general, TDS was significantly higher in the KERN study unit than in the SESJ study unit (table 5). The higher concentrations of TDS in the KERN study unit may be affected by sediments from the Tehachapi and San Emigdio Mountains and the Coast Ranges which contain marine deposits. TDS was negatively correlated with pH in the study units (table 11). This relation also was found in two GAMA study units in the northern part of the San Joaquin Valley (Bennett and others, 2010; Landon and others, 2010a). TDS also was positively correlated to agricultural land use in the KERN study unit. TDS was correlated in orchard and vineyard land use (a subset of agricultural land use) and well depth in the SESJ study unit if the domestic wells are included in the analysis. Agricultural irrigation is a potential source of TDS. TDS also was negatively correlated to the number of septic systems in the KERN study unit. On the basis of these relations, higher concentrations of TDS in SESJ primarily may be a result of human activities, while higher concentrations of TDS in KERN primarily may be from natural sources as well as human activities.

Sulfate

Sulfur occurs naturally in both igneous and sedimentary rocks as metallic sulfides. Pyrite crystals that occur in many sedimentary rocks are a major source of both ferrous iron and sulfate in groundwater (Hem, 1985). Sulfate also occurs in evaporate sediments such as gypsum (calcium sulfate). Sulfur also is applied as an agricultural fertilizer on parts of the San Joaquin Valley (Jurgens and others, 2008).

Sulfate was detected at high relative-concentrations in 8.3 percent and at moderate relative-concentrations in 6.2 percent of the primary aquifers in the KERN study unit (table 8; fig. 16C). Sulfate was detected at moderate relative-concentrations in 1.7 percent of the primary aquifers in the SESJ unit (table 8). The high relative-concentrations of sulfate primarily occurred in CDPH-grid and CDPH-other wells in the area south of the city of Bakersfield (fig. 17K) near the Tehachapi and San Emigdio Mountains.

Similar to TDS, sulfate concentrations were higher in the KERN study unit than in the SESJ study unit (table 5). High concentrations of sulfate may be from sediments from the Tehachapi and San Emigdio Mountains. Sulfate was positively correlated to agricultural land use in the KERN study unit but not in the SESJ study unit (table 11). However, sulfate was correlated with orchard and vineyard land use (a subset of agricultural land-use) in the SESJ study unit if the domestic wells are included in the analysis. Sulfate concentrations also decreased with well depth in the SESJ study unit. Sulfate was significantly higher in mixed-age groundwater than in pre-modern-age groundwater in the SESJ study unit, but sulfate concentrations were not significantly different between modern-age water and mixed or pre-modern age groundwater. This relation may be affected by well depth. These relations suggest elevated sulfate concentrations in SESJ primarily are from anthropogenic sources, while the high sulfate concentrations in the southern part of KERN are from natural sources.

Chloride

Chloride most commonly is associated with sedimentary rocks, particularly evaporates. Where porous rocks are submerged by the sea, soluble salts infiltrate the rock. Fine-grained marine shales might retain chloride for long periods of time (Hem, 1985). Chloride is present in precipitation as a result of entrainment of marine salts into the air at the ocean's surface. Human activities also may be a source of chloride in some areas.

Chloride was detected at high relative-concentrations in 2.1 percent and at moderate relative-concentrations in 4.2 percent of the primary aquifers in the KERN study unit (table 8; fig. 16C). Chloride was detected in high relative-concentrations in 0.3 percent in the SESJ study unit (table 8).

Similar to TDS, chloride concentration was higher in the KERN study unit than in the SESJ study unit (table 5). Chloride was positively correlated with depth to top-of-perforations and orchard and vineyard land use in the KERN study unit (table 11). Chloride concentration was significantly higher in the distal part of the SESJ study unit than in the eastern part. These relations suggest chloride concentrations are affected by both human activities and natural sources similar to sulfate.

Organic Constituents

Organic and special-interest constituents, unlike inorganic constituents, usually are of anthropogenic origin. VOCs may be present in paints, solvents, fuels, refrigerants, can be byproducts of water disinfection, and are characterized by their tendency to evaporate. In this report, VOCs are classified as THMs, solvents, and other VOCs. Pesticides are used to control weeds, insects, or fungi in agricultural, urban, and suburban settings. Pesticides are classified as fumigants and pesticides.

Organic constituents with human-health benchmarks were detected at high relative-concentrations in 4.8 percent and 2.1 percent of the primary aquifers in the SESJ and KERN study units, respectively (table 12). Benzene, 1,2-dibromo-3-chloropropane (DBCP), carbon tetrachloride, tetrachloroethene (PCE), and trichloroethene (TCE) were the organic constituents detected at high relative-concentrations in grid wells (table 9). Organic constituents with human-health benchmarks were detected at moderate relative-concentrations (greater than 0.1 but less than or equal to 1.0) in 11 percent and 8.5 percent of the primary aquifers in the SESJ and KERN study units, respectively (table 12). Organic constituents with moderate relative-concentrations in either the SESJ or KERN study units were the solvents dichloromethane and 1,2-dichloroethane (1,2-DCA), the pesticide dieldrin, and the fumigant 1,2-dibromoethane (EDB). All of the other organic constituents with human-health benchmarks were detected at low relative-concentrations (less than or equal to 0.1) or were not detected (fig. 22). In addition, several organic constituents—the THM chloroform, the pesticides atrazine, simazine, dinoseb, and bromacil, the fumigants 1,2,3-trichloropropane (1,2,3-TCP) and 1,2-dichloropropane (1,2-DCP)—were prevalent (detection frequency greater than 10 percent in USGS-grid wells) in the primary aquifers (fig. 22). The relative-concentrations of selected organic compounds are shown in figure 23 in relations to the study areas of the SESJ study unit or KERN study unit in which they are detected.

Table 12. Aquifer-scale proportions for organic constituent classes and constituents of special interest for the two southern San Joaquin Valley study units, California GAMA Priority Basin Project.

[Aquifer-scale proportions are given in percentage of area of the primary aquifer. All values greater than 10 percent are rounded to the nearest 1 percent; values less than 10 percent are rounded to the nearest 0.1 percent; values may not add up to 100 percent because of rounding. THMs, trihalomethanes; VOCs, volatile organic compounds; NDMA, N-nitrosodimethylamine]

Constituent	Aquifer-scale proportion (percent)		
	High	Moderate	Low or not detected
Southeast San Joaquin Valley study unit			
Organics with human-health benchmarks			
THMs	0.0	1.2	99
Solvents	0.7	1.2	88
Other VOCs	1.2	0.0	99
Fumigants	3.6	9.6	86
Pesticides	0.0	2.4	98
Any organic constituent	4.8	11	84
Constituents of special interest			
Perchlorate, NDMA	1.2	19	80
Kern County Subbasin study unit			
Organics with human-health benchmarks			
THMs	0.0	4.2	96
Solvents	0.1	4.3	95
Other VOCs	0.0	2.1	98
Fumigants	2.1	4.3	93
Pesticides	0.0	0.0	100
Any organic constituent	2.1	8.5	89
Constituents of special interest			
Perchlorate	0.0	6.4	94

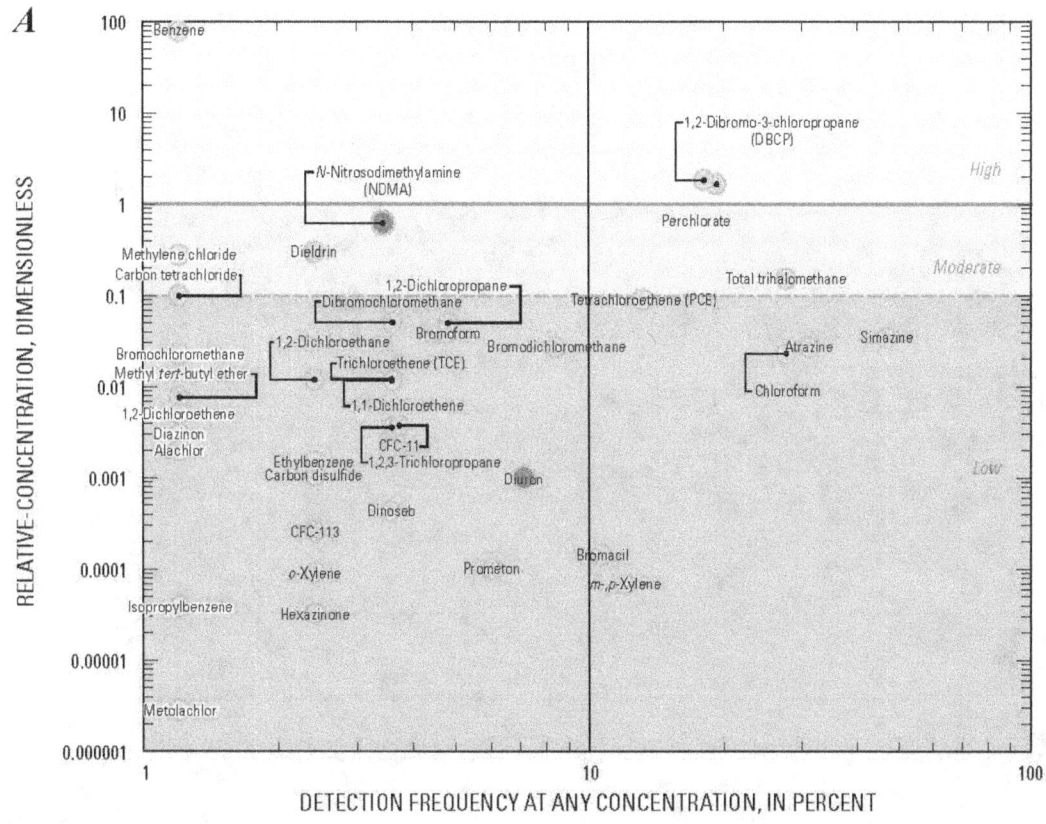

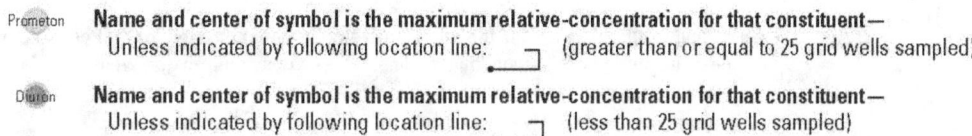

EXPLANATION

Prometon — **Name and center of symbol is the maximum relative-concentration for that constituent—**
Unless indicated by following location line: ⌐ (greater than or equal to 25 grid wells sampled)

Diuron — **Name and center of symbol is the maximum relative-concentration for that constituent—**
Unless indicated by following location line: ⌐ (less than 25 grid wells sampled)

Figure 22. Detection frequency and maximum relative-concentration for organic and special-interest constituents detected in USGS-grid wells in (*A*) the Southeast San Joaquin Valley study unit and (*B*) the Kern County Subbasin study unit., California GAMA Priority Basin Project.

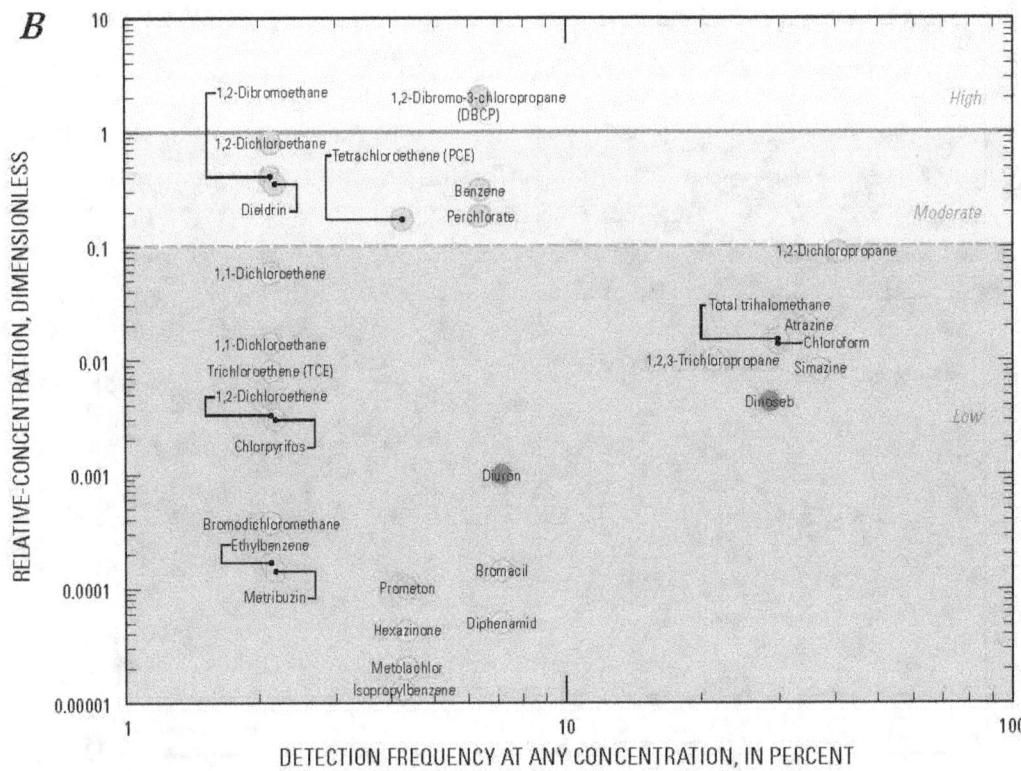

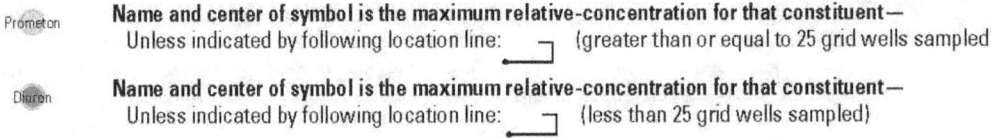

EXPLANATION

Prometon   **Name and center of symbol is the maximum relative-concentration for that constituent—**
Unless indicated by following location line: ⌐ (greater than or equal to 25 grid wells sampled)

Diuron   **Name and center of symbol is the maximum relative-concentration for that constituent—**
Unless indicated by following location line: ⌐ (less than 25 grid wells sampled)

Figure 22.—Continued

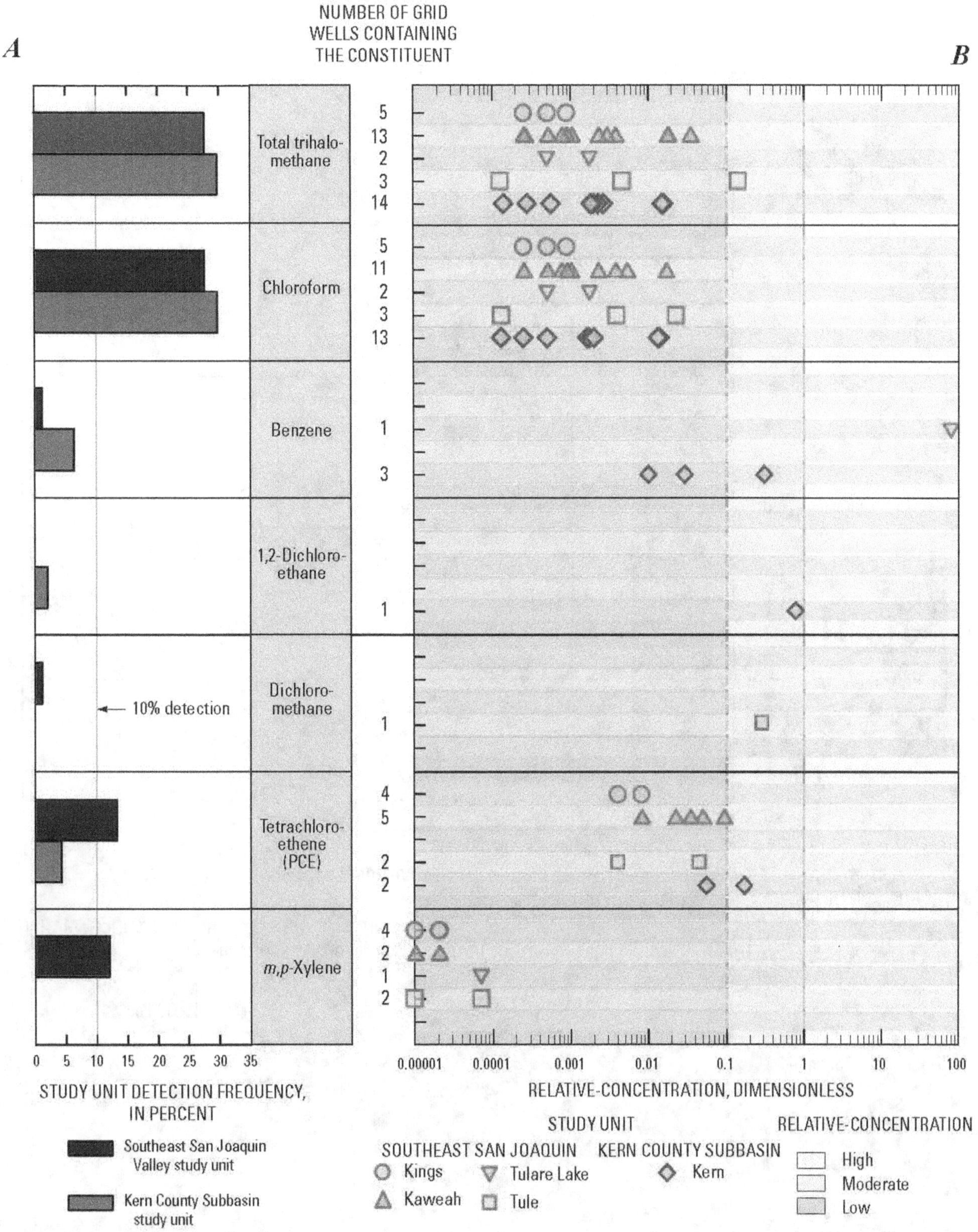

NUMBER OF GRID
WELLS CONTAINING
THE CONSTITUENT

A

B

STUDY UNIT DETECTION FREQUENCY,
IN PERCENT

RELATIVE-CONCENTRATION, DIMENSIONLESS

Southeast San Joaquin
Valley study unit

Kern County Subbasin
study unit

STUDY UNIT

SOUTHEAST SAN JOAQUIN   KERN COUNTY SUBBASIN

○ Kings   ▽ Tulare Lake   ◇ Kern

△ Kaweah   ☐ Tule

RELATIVE-CONCENTRATION

High
Moderate
Low

Figure 23. (*A–H*) Detection frequencies and relative-concentrations of selected organic and special-interest constituents in USGS-grid wells in the two southern San Joaquin Valley study units, California GAMA Priority Basin Project.

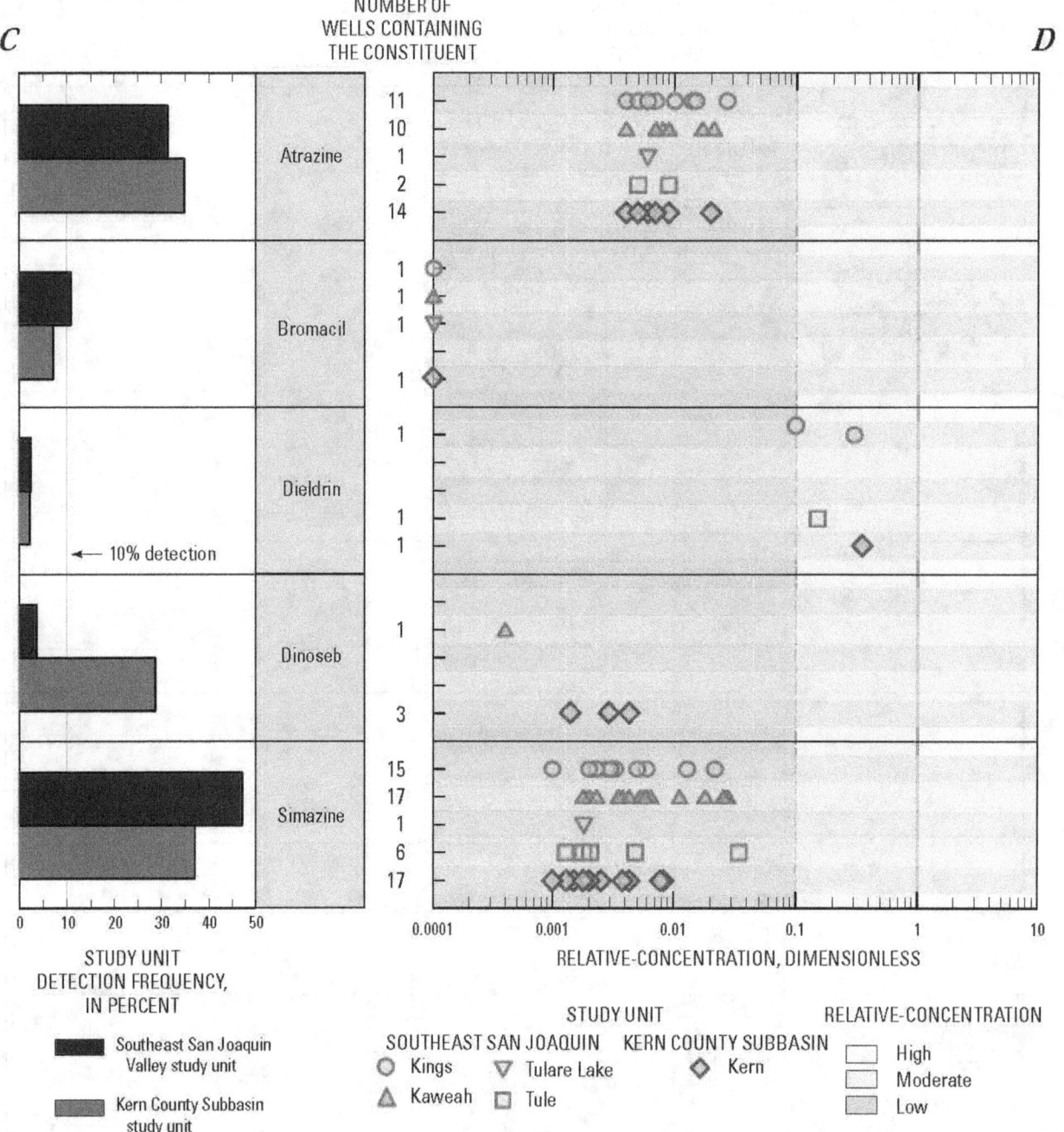

Figure 23.—Continued

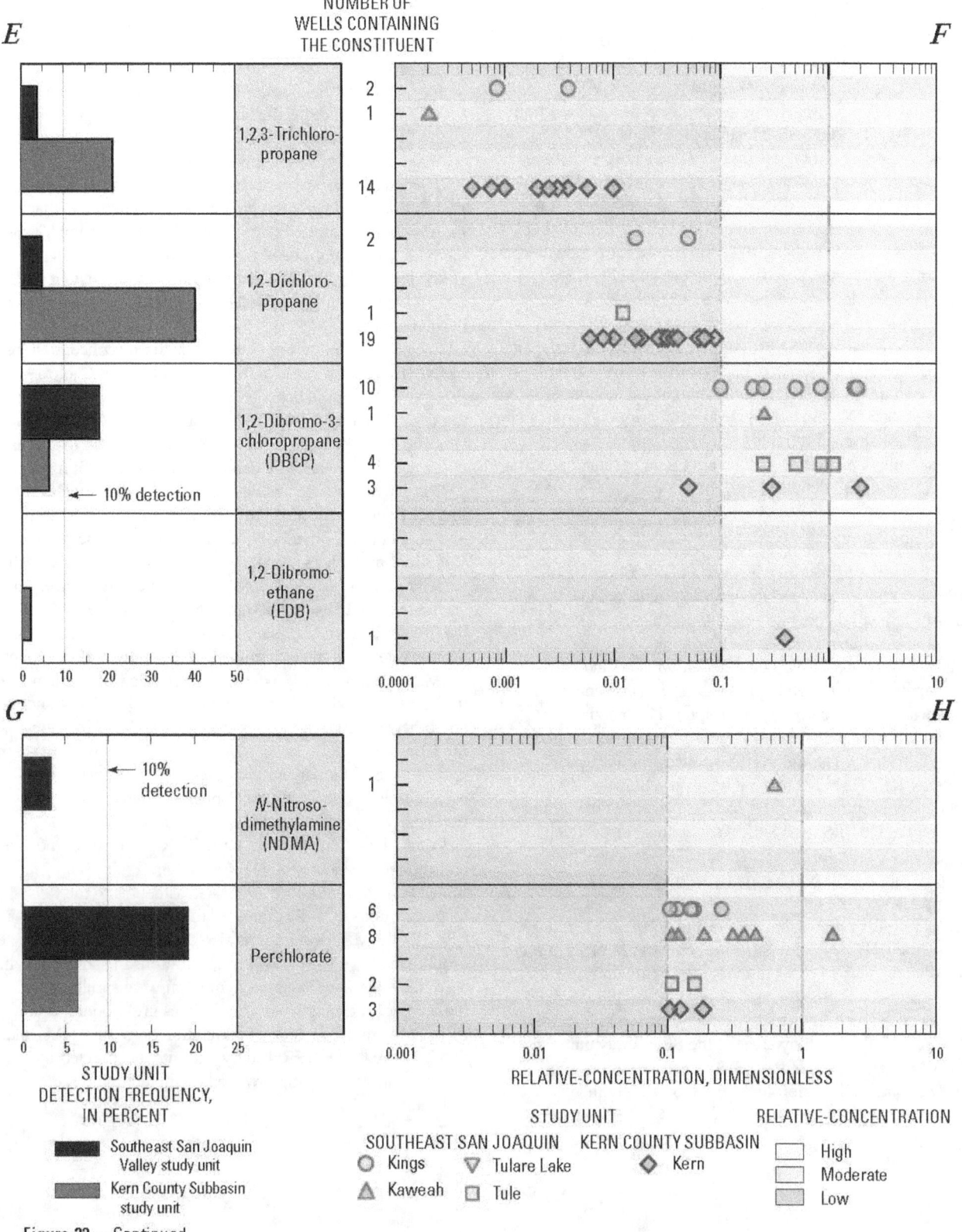

NUMBER OF
WELLS CONTAINING
THE CONSTITUENT

Figure 23.—Continued

Volatile Organic Compounds

VOCs discussed in this report are classified as THMs, solvents, and other VOCs. More than one VOC was detected in almost one-half of the USGS-grid wells with VOC detections. Figure 24A shows the number of VOC detections in USGS-grid and USGS-understanding wells and CDPH wells. Wells with more than one VOC generally were located in the eastern part of the study units. The number of VOC detections decreased with well depth and depth to top-of-perforations in the SESJ study unit (table 13). The number of VOC detections was higher in modern-age groundwater than in groundwater of pre-modern-age in the SESJ study unit. These relations were not observed in the KERN study unit. The lack of a relation with depth in the KERN study unit may be because most wells in the KERN study unit are deep and do not span the range of depths as the wells in the SESJ study unit do (table 5). The correlation of the number of VOCs detected with DO and normalized lateral position was positive in the SESJ study unit (table 13). The relation with lateral position may be affected by the correlation of lateral position with DO and well depth (table 6B).

Trihalomethanes

Water used for drinking water and other household uses in both domestic and municipal systems commonly is disinfected with hypochlorite solutions (bleach). As a side effect of disinfection, the hypochlorite reacts with organic matter to produce THMs and other chlorinated and/or brominated disinfection byproducts. The THMs analyzed in this study were chloroform, bromodichloromethane, dibromochloromethane, and bromoform. THMs, as a class, were not detected at high relative-concentrations in the SESJ or KERN study units but were detected at moderate relative-concentrations in 1.2 and 4.2 percent of the primary aquifers in the SESJ and KERN study units, respectively (table 12). Chloroform was the only THM detected in 10 percent or more of the primary aquifers (fig. 22). Comparison of the relative-concentrations for total THMs and chloroform shows that chloroform accounted for most of the total THMs in almost all the samples (22 of 23 in the SESJ study unit; 14 of 14 in the KERN study unit; Burton and others, 2008; Shelton and others, 2008). Chloroform was detected in more than 25 percent of the samples and was detected in all four study areas of the SESJ study unit and in the KERN study unit at low relative-concentrations (figs. 23A and B). Nationally, chloroform was the most frequently

detected VOC in aquifers in studies conducted by the USGS National Water-Quality Assessment (NAWQA) Program (Zogorski and others, 2006).

Correlations were done using total THMs as the water-quality variable because the MCL (80 µg/L) is for total THMs and not for the individual THM compounds. In the SESJ study unit, THMs were positively correlated with DO, lateral position, number of septic systems, and LUFT density; negatively correlated with depth to top-of-perforations; and higher in modern-age groundwater than in pre-modern-age groundwater (tables 6 and 13). Total THMs were negatively correlated to natural land use in the KERN study unit (table 13). These correlations are consistent with an anthropogenic origin for the THMs detected in groundwater in these areas.

THM concentrations were low with low concentrations of DO (less than 4 mg/L) in the SESJ study unit (table 13; fig. 25). The relation between THMs and DO could indicate the degradation of THMs in increasingly anoxic groundwater (Pavelic and others, 2006) or simply may indicate that both constituents decrease with depth. Positive correlations of THMs and DO also have been noted in the Priority Basin Project Central Eastside study unit north of the SESJ study unit (Landon and others, 2010a) and in nationwide analysis (Squillace and others, 2004; Zogorski and others, 2006). The correlation of THMs with lateral position could be affected by the relation between DO and lateral position as well as by the greater density of potential sources of THMs, including septic systems, in the eastern part of the study unit. Although THMs were not directly positively correlated with urban land use, as was found by Bennett and others (2010) and Landon and others (2010a) farther north in the San Joaquin Valley, the increase in septic system and LUFT density in the eastern part of the SESJ study unit may represent greater urban activities and concentration of potential sources of THMs in that area (table 5). However, the negative correlation of THMs with natural land use in the KERN study unit indicates anthropogenic sources of THMs.

THM concentrations decreased as depths to the top-of-perforations in wells increased in the SESJ study unit (fig. 25). THM concentrations also decreased in older groundwater (table 5). THM concentrations in samples having a groundwater age classified as modern were significantly higher than in groundwater classified as pre-modern in the SESJ study unit. The lack of correlation between THMs and depth or age in the KERN study unit may be affected by the lack of relatively shallow wells sampled in the KERN study unit.

Table 13. Results of non-parametric correlation analysis (Spearman's method) between selected organic and special-interest water-quality constituents and potential explanatory factors in the two southern San Joaquin Valley study units, California GAMA Priority Basin Project.

[Aquifer-scale proportions are from the grid-based method unless otherwise stated. Spearman's correlation statistic (ρ) for a significant correlation on the basis of significance level (p) less than threshold value (α) of 0.05. **Abbreviations:** LUFTs, leaking underground fuel tanks; VOC, volatile organic compound; THMs, trihalomethanes; DBCP, 1,2-Dibromo-3-chloropropane; MCL-US, USEPA maximum contaminant level; ns, no significant correlation; <, less than; GAMA, Groundwater Ambient Monitoring and Assessment Program; USEPA, U.S. Environmental Protection Agency]

Constituent, constituent class, or special-interest constituent	Benchmark type	High aquifer proportion (percent)	Grid and understanding wells used in analysis							Grid wells used in analysis			
			Depth to top-of-perforations	Well depth	pH	Dissolved oxygen	LUFTs	Number of septic tanks or cesspools[1]	Orchard and vineyard land use[1]	Agricultural land use[1] (percent)	Urban land use[1] (percent)	Natural land use[1] (percent)	Normalized lateral position from valley trough
			ρ: Spearman's correlation statistic/p: significance level										
Southeast San Joaquin study unit													
Number of VOC detections[2]	variable	2.4	-0.312 / 0.007	-0.234 / 0.034	ns	0.460 / <0.001	ns	ns	ns	ns	ns	ns	0.302 / 0.006
THMs, total[3]	MCL-US	0.0	-0.263 / 0.025	ns	ns	0.479 / <0.001	0.213 / 0.034	0.208 / 0.039	ns	ns	ns	ns	0.259 / 0.018
Number of solvent detections[3]	variable	0.7	-0.301 / 0.010	ns	ns	0.271 / 0.007	ns	0.275 / 0.006	ns	-0.287 / 0.009	0.287 / 0.009	ns	0.255 / 0.020
Number of other VOC detections[3]	variable	1.2	ns	ns	-0.206 / 0.041	ns	ns	ns	ns	ns	ns	ns	ns
DBCP	MCL-US	3.6	ns	ns	-0.349 / 0.004	0.400 / <0.001	ns	ns	ns	ns	ns	ns	0.280 / 0.010
Number of pesticide detections[3]	variable	0.0	-0.521 / <0.001	-0.474 / <0.001	-0.360 / 0.003	0.408 / <0.001	ns	0.249 / 0.013	ns	-0.232 / 0.035	0.246 / 0.025	ns	0.475 / <0.001
Perchlorate	MCL-US	1.2	-0.366 / 0.002	-0.377 / 0.001	-0.335 / 0.006	0.505 / <0.001	0.283 / 0.005	0.282 / 0.005	ns	ns	ns	ns	0.389 / <0.001
Kern County Subbasin study unit													
Number of VOC detections[2]	variable	0.1	ns	ns	ns	ns	ns	ns	ns	ns	-0.318 / 0.030	ns	ns
THMs, total[3]	MCL-US	0.0	ns	ns	ns	ns	ns	ns	ns	ns	ns	-0.311 / 0.0338	ns
Number of solvent detections	variable	0.1	ns	ns	ns	ns	ns	0.305 / 0.031	-0.290 / 0.041	ns	ns	ns	ns
Number of other VOC detections	variable	0.0	ns	ns	ns	ns	-0.360 / 0.011	ns	ns	ns	ns	ns	ns
DBCP	MCL-US	2.1	ns	ns	ns	ns	ns	ns	ns	ns	ns	ns	ns
Number of pesticide detections[3]	variable	0.0	ns	-0.300 / 0.039	ns	ns	ns	ns	ns	ns	ns	ns	ns
Perchlorate	MCL-US	0.0	ns	ns	ns	0.410 / 0.007	ns	ns	ns	ns	ns	ns	ns

Table 13. Results of non-parametric correlation analysis (Spearman's method) between selected organic and special interest water-quality constituents and potential explanatory factors in the two southern San Joaquin Valley study units, California GAMA Priority Basin Project.

[Aquifer-scale proportions are from the grid-based method unless otherwise stated. Spearman's correlation statistic (ρ) for a significant correlation on the basis of significance level (p) less than threshold value (α) of 0.05. **Abbreviations:** LUFTs, leaky underground fuel tanks; VOC, volatile organic compound; THMs, trihalomethanes; DBCP, 1,2-Dibromo-3-chloropropane; MCL-US, USEPA maximum contaminant level; ns, no significant correlation; <, less than; GAMA, Groundwater Ambient Monitoring and Assessment Program; USEPA, U.S. Environmental Protection Agency]

Constituent, constituent class, or special-interest constituent	Benchmark type	High aquifer proportion (percent)	Grid and understanding wells used in analysis							Grid wells used in analysis			
			Depth to top-of-perforations	Well depth	pH	Dissolved oxygen	LUFTs	Number of septic tanks or cesspools[1]	Orchard and vineyard land use[1]	Agricultural land use[1] (percent)	Urban land use[1] (percent)	Natural land use[1] (percent)	Normalized lateral position from valley trough
							ρ: Spearman's correlation statistic/p: significance level						
						Southeast San Joaquin Valley study unit with Domestic Wells included							
DBCP	MCL-US			ns	-0.349 0.004			ns	0.155 0.016				
Perchlorate	MCL-US			-0.387 <0.001	-0.335 0.006			ns	ns				

[1] Land use percentages and number of septic tanks are within a radius of 500 meters around each well included in analysis.

[2] Does not include VOCs classified as fumigants.

[3] Classes of compounds that include constituents with high values less than 2 percent, moderate values, or detection frequencies at any concentration greater than 10 percent.

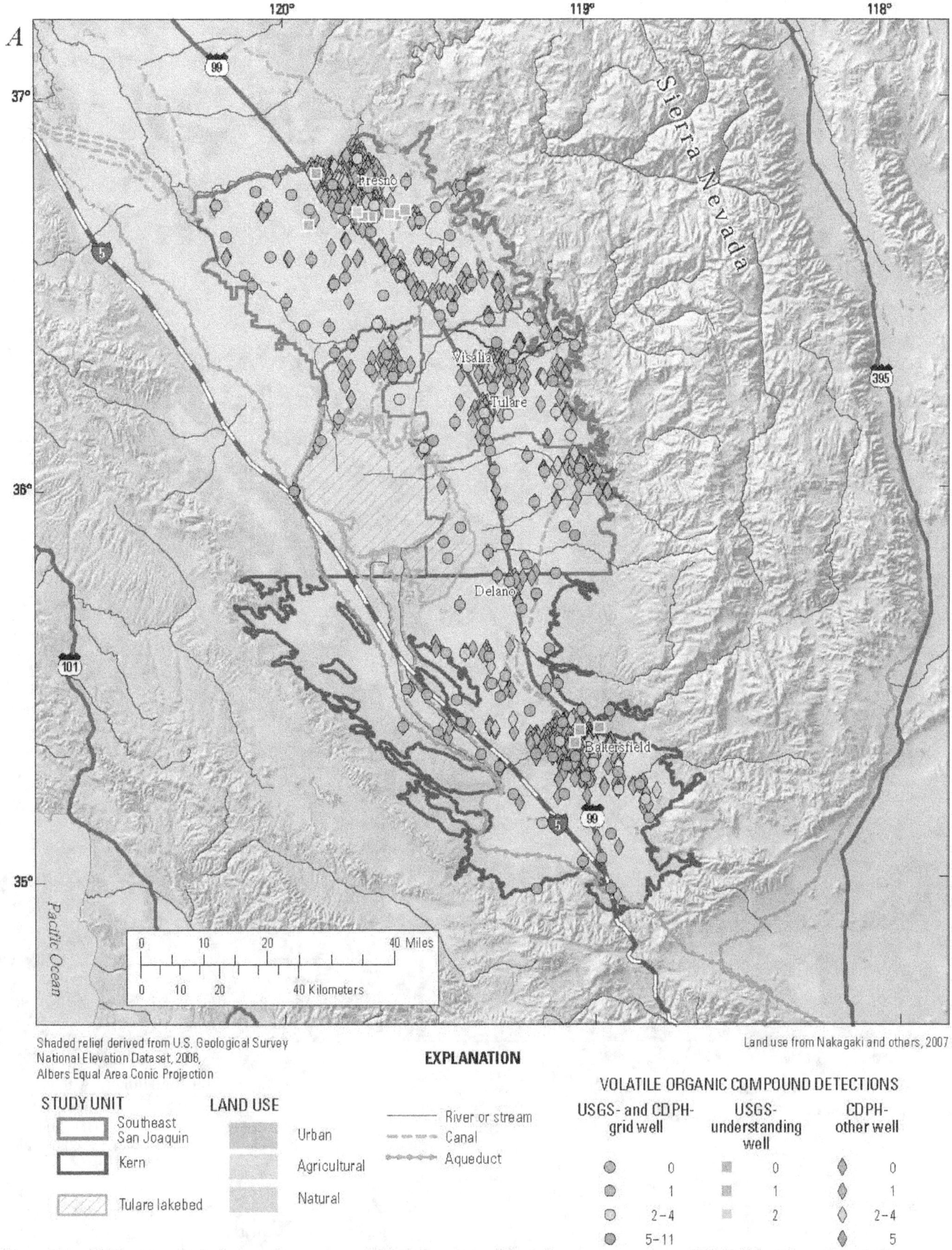

Figure 24. (A) Number of volatile organic compound (VOC) detections, (B) relative-concentrations of DBCP, (C) number of pesticide detections, and (D) relative-concentration of perchlorate in USGS-grid wells, USGS-understanding wells, and CDPH-other wells in the two southern San Joaquin Valley study units, California GAMA Priority Basin Project.

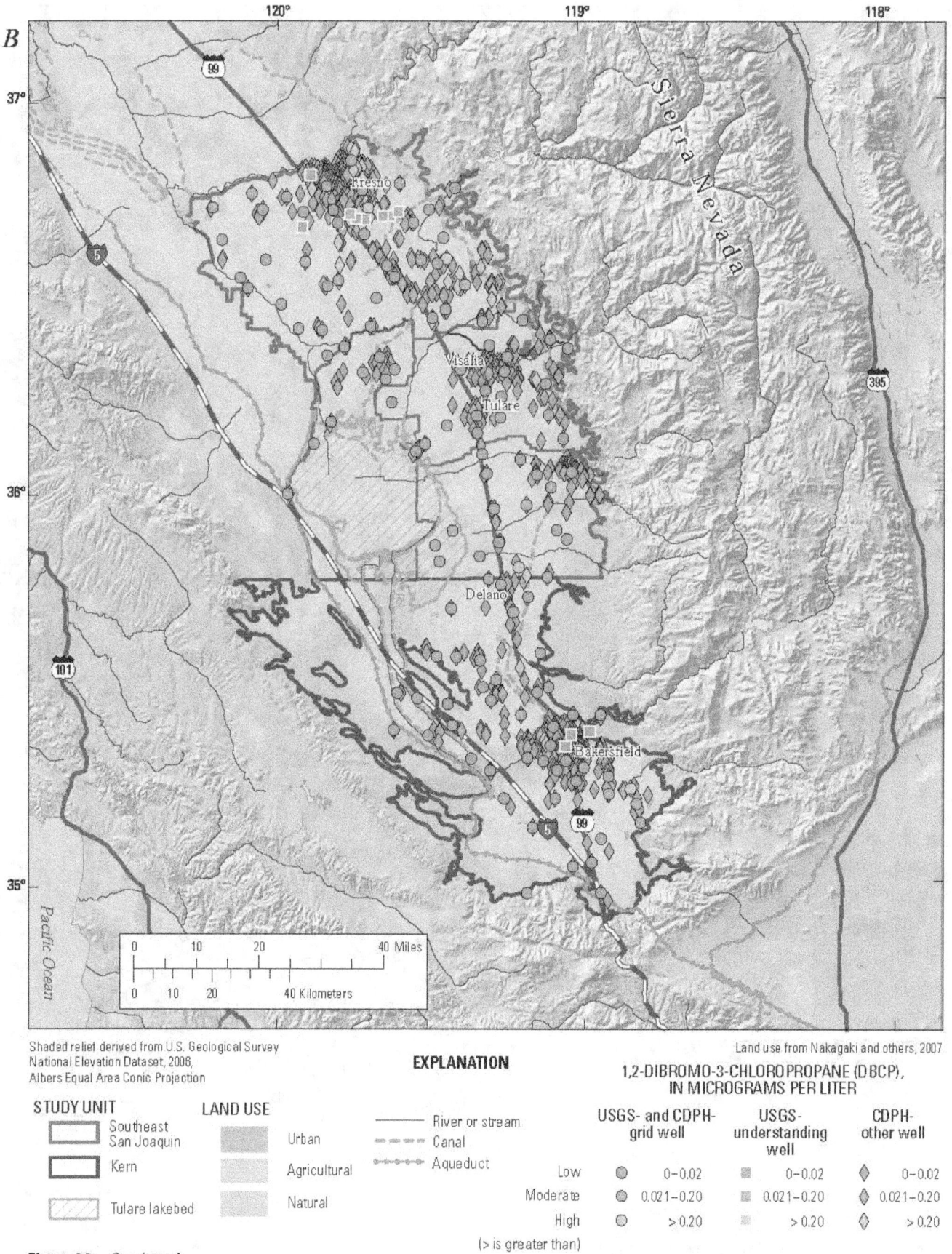

Shaded relief derived from U.S. Geological Survey
National Elevation Dataset, 2008,
Albers Equal Area Conic Projection

EXPLANATION

Land use from Nakagaki and others, 2007

**1,2-DIBROMO-3-CHLOROPROPANE (DBCP),
IN MICROGRAMS PER LITER**

STUDY UNIT

☐ Southeast San Joaquin

☐ Kern

▨ Tulare lakebed

LAND USE

Urban

Agricultural

Natural

—— River or stream

– – – Canal

⊷⊷⊷ Aqueduct

	USGS- and CDPH-grid well	USGS-understanding well	CDPH-other well
Low	● 0–0.02	▪ 0–0.02	◆ 0–0.02
Moderate	● 0.021–0.20	▪ 0.021–0.20	◆ 0.021–0.20
High	● >0.20	▪ >0.20	◆ >0.20

(> is greater than)

Figure 24.—Continued

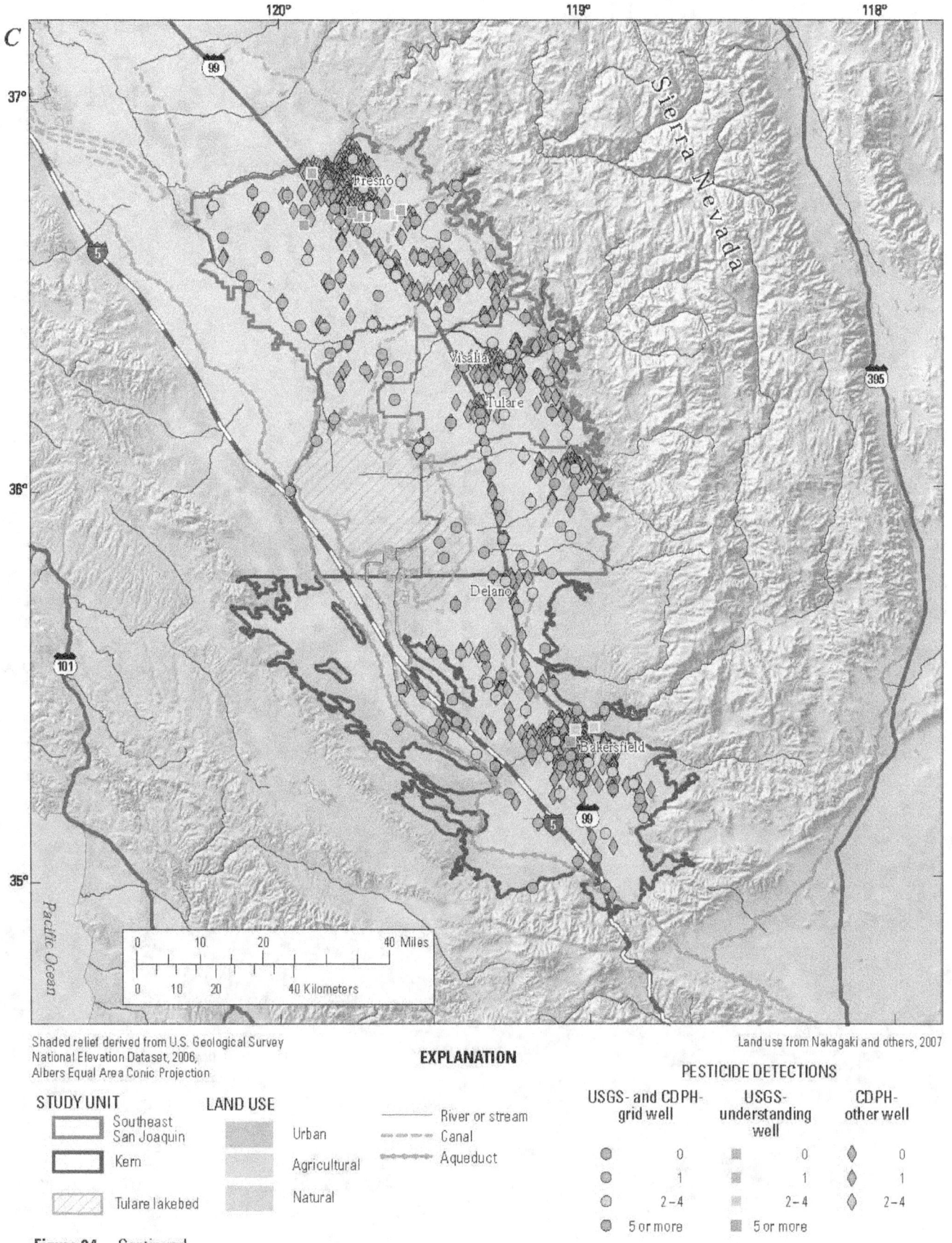

Shaded relief derived from U.S. Geological Survey
National Elevation Dataset, 2006,
Albers Equal Area Conic Projection

Land use from Nakagaki and others, 2007

EXPLANATION

PESTICIDE DETECTIONS

STUDY UNIT

☐ Southeast San Joaquin

☐ Kern

▨ Tulare lakebed

LAND USE

Urban

Agricultural

Natural

——— River or stream

⎯ ⎯ Canal

●—●—● Aqueduct

USGS- and CDPH-grid well

● 0
● 1
● 2 – 4
● 5 or more

USGS-understanding well

■ 0
■ 1
■ 2 – 4
■ 5 or more

CDPH-other well

◆ 0
◆ 1
◆ 2 – 4

Figure 24.—Continued

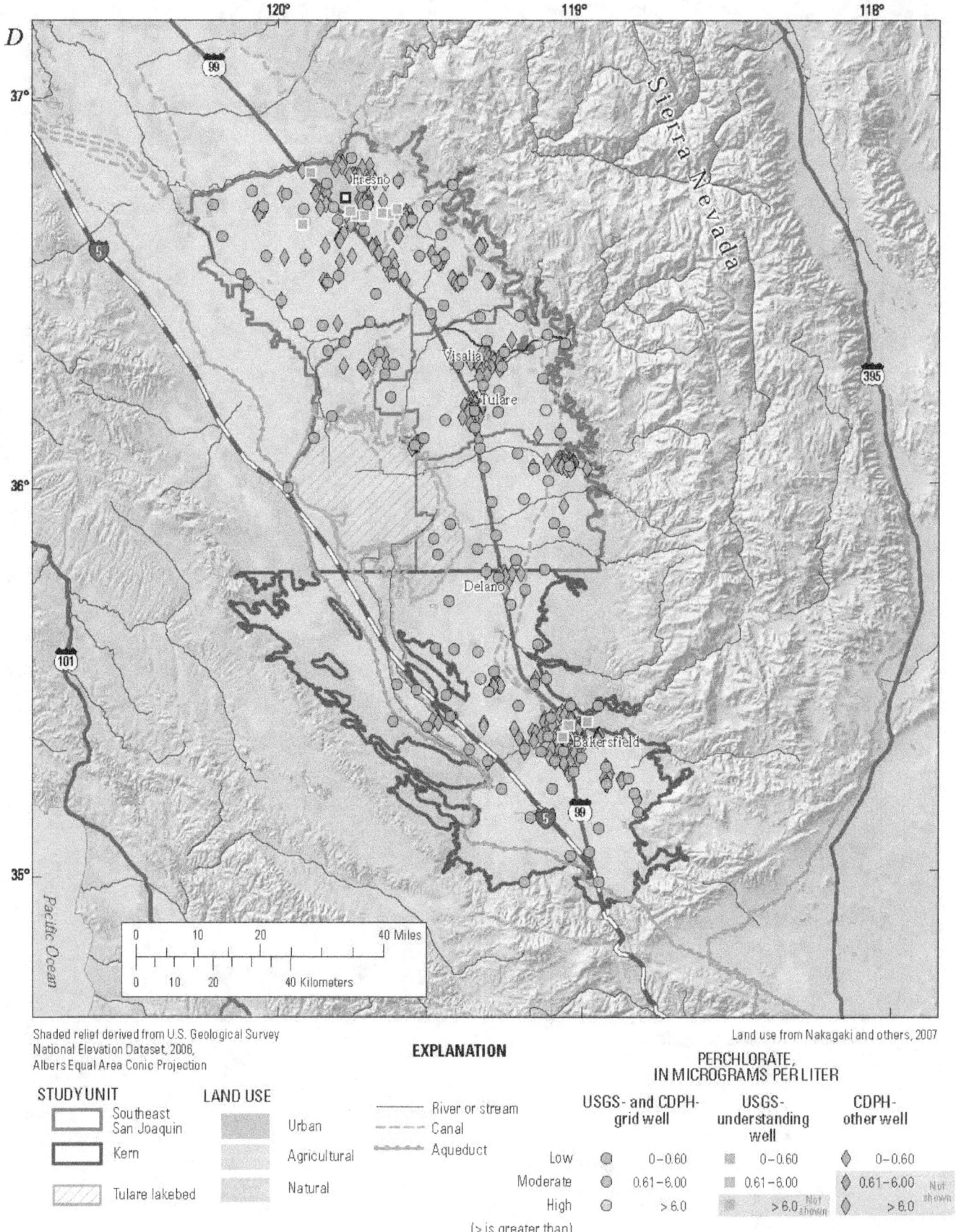

Shaded relief derived from U.S. Geological Survey
National Elevation Dataset, 2006,
Albers Equal Area Conic Projection

Land use from Nakagaki and others, 2007

EXPLANATION

**PERCHLORATE,
IN MICROGRAMS PER LITER**

STUDY UNIT

	Southeast San Joaquin
	Kern
	Tulare lakebed

LAND USE

	Urban
	Agricultural
	Natural

— River or stream
– – Canal
~~~ Aqueduct

USGS- and CDPH-grid well

Low	0–0.60
Moderate	0.61–6.00
High	> 6.0

USGS-understanding well

0–0.60
0.61–6.00
> 6.0 Not shown

CDPH-other well

0–0.60
0.61–6.00 Not shown
> 6.0

(> is greater than)

Figure 24.—Continued

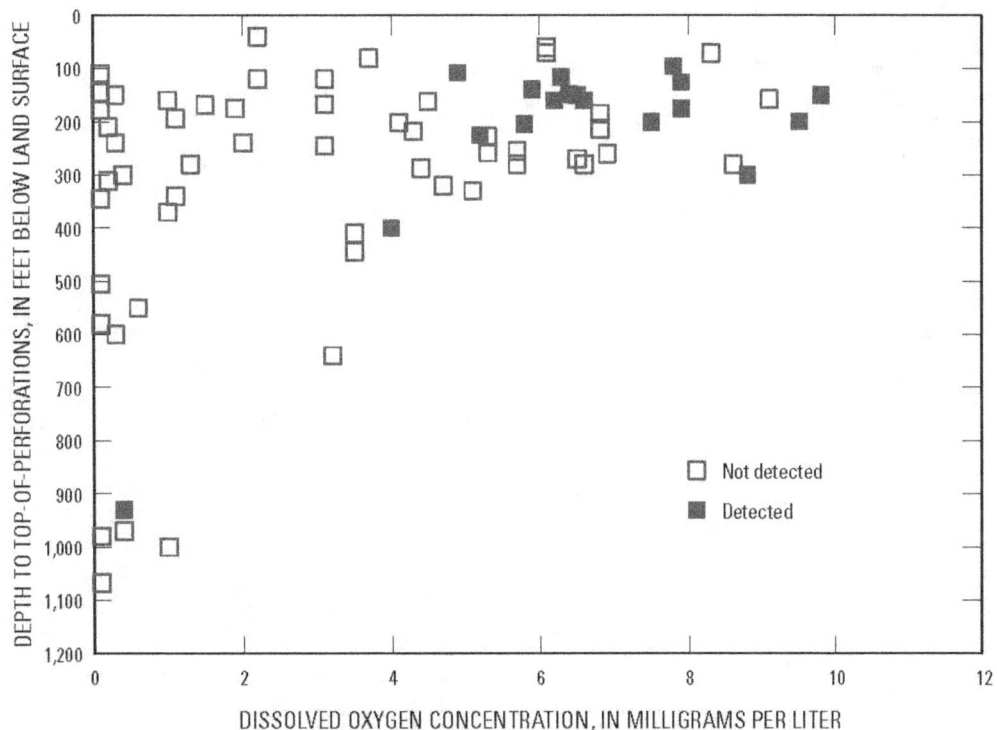

Figure 25. Relation of detections of total trihalomethanes (THMs) to depth to top-of-perforations and dissolved oxygen for USGS-grid and USGS-understanding wells in the Southern San Joaquin Valley study units, California GAMA Priority Basin Project.

Solvents

Solvents are used for various industrial, commercial, and domestic purposes. Solvents, as a class, were detected at high relative-concentrations in 0.7 and 0.1 percent, at moderate relative-concentrations in 1.2 and 4.3 percent, and at low relative-concentrations or not detected in 88 and 95 percent of the primary aquifers in the SESJ and KERN study units, respectively (table 12). Carbon tetrachloride, TCE, and PCE were detected at high relative-concentrations in less than 1 percent of the primary aquifers in the SESJ study unit (table 9). Carbon tetrachloride and TCE were detected at high relative-concentrations in less than 1 percent of the primary aquifers in the KERN study unit (table 9). Dichloromethane was detected at moderate relative-concentrations in the SESJ study unit; PCE and 1,2-DCA were each detected at moderate relative-concentrations in the KERN study unit (fig. 23B).

The number of solvent detections was positively correlated with percentage of urban land use, DO, and number of septic systems in the SESJ study unit (table 13). The number of solvent detections also was higher in the eastern part of the SESJ study unit where most of the urban land use and septic systems are located than in the western part of the SESJ study unit. The number of solvent detections also was positively correlated to septic systems in the KERN study unit. The number of solvent detections was negatively correlated

to agricultural land use and to depth to the top-of-perforations (table 5), and solvent detections were significantly greater in modern than in pre-modern aged groundwater in the SESJ study unit (table 5). Solvent detections were negatively correlated with orchard and vineyard land use in the KERN study unit (table 13).

The number of solvents detected was greater in wells when the percentage of urban land use was greater than 40 percent (7 out of 9 wells with detections, fig. 26) in the SESJ study unit. Correlations of solvents with urban land use also were found in two other GAMA study units in the San Joaquin Valley (Bennett and others, 2010; Landon and others, 2010a). Nationally, solvent concentrations also have been correlated strongly with percentage of urban land use because most solvents are of anthropogenic origin (Zogorski and others, 2006; Moran and others, 2007). A previous investigation in the Fresno area showed that urban land use was the best predictor for the detection of VOCs (Wright and others, 2004). The correlation of solvents with the density of septic systems likely is a result of the positive correlation of urban land use with septic systems (table 6B). The low number of solvent detections in the KERN study unit makes it difficult to identify relations; the negative correlation of solvent detections with orchard and vineyard land use may indicate other factors.

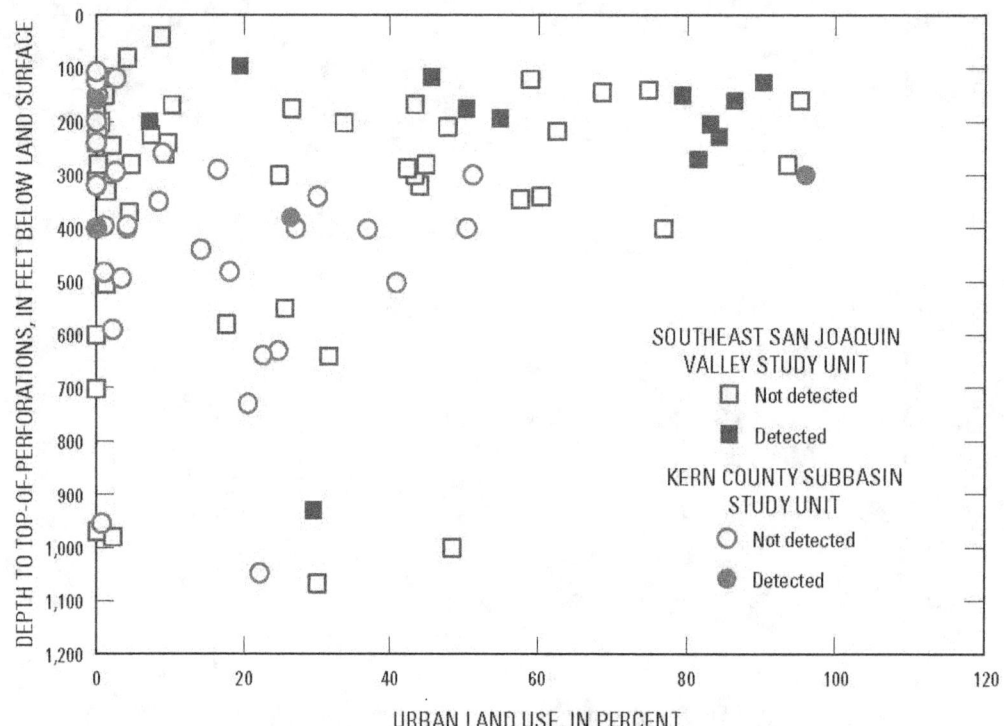

Figure 26. Relation of solvent detection with depth to top-of-perforation and percentage of urban land use for USGS-grid and USGS-understanding well samples in the two southern San Joaquin Valley study units, California GAMA Priority Basin Project.

Solvent detections decreased as depth to the top-of-perforation in wells increased in the southern San Joaquin Valley. Almost all detections (14 of 17) were in wells where the depth to top-of-perforations was less than 300 ft (fig. 26). Solvent detections also decreased in older groundwater (table 5). These relations are consistent with the results of two other GAMA study units in the San Joaquin Valley (Bennett and others, 2010; Landon and others, 2010a).

Other Volatile Organic Compounds

Other VOCs, as a class, were detected at high relative-concentrations in 1.2 percent of the primary aquifers in the SESJ study unit and not in the KERN study unit (table 12). Other VOCs were detected at moderate relative-concentrations in 2.1 percent of the primary aquifers in the KERN study unit, and at low relative-concentrations or not detected in 99 percent and 98 percent of the primary aquifers in the SESJ and KERN study units, respectively. Benzene, was detected at high relative-concentrations in one well in the Tulare Lake study area (fig. 23B). Benzene was detected at moderate and at low relative-concentrations in three wells in KERN.

Detections of other VOCs consisted of gasoline hydrocarbons, compounds used for organic synthesis, and two refrigerants (Burton and others, 2008; Shelton and others, 2008). Detections of other VOCs were negatively correlated with DO in the SESJ study unit and with LUFTs in the KERN study (table 13). Other VOC detections were greater in pre-modern than in modern-age groundwater in the KERN study area (table 6).

The negative correlation with LUFTs was unexpected because most of the detections in the KERN study unit were gasoline hydrocarbons. However, the higher detection frequency in pre-modern-age groundwater than in modern-age groundwater suggests the source of the gasoline hydrocarbons is natural from deep in the aquifer. There are several oil and gas fields located near the detections of gasoline hydrocarbons. It is uncertain if the correlation of other VOCs with DO in the SESJ study unit is an explanatory factor. Almost all of the detections of other VOCs in the SESJ study unit are in the eastern half of the study unit where DO is higher. Because detection frequencies of these compounds are low, conclusions based on these data are uncertain. Other VOC detections were not correlated to any other explanatory factor in either study unit.

Fumigants

Ten VOCs used primarily as fumigants to control pests in agriculture and in households, or synthesis byproducts included in fumigant mixtures, were grouped into the constituent class of fumigants. The classification of nine of these constituents as fumigants was determined by the USGS NAWQA Program (Zogorski and others, 2006). 1,2,3-trichloropropane (1,2,3-TCP) is classified as having a primary use as a solvent and in the synthesis of some organic compounds (Zogorski and others, 2006), but 1,2,3-trichloropropane (1,2,3-TCP) also was a synthesis byproduct in fumigant mixtures in use from the 1950s until the early 1980s (Oki and Giambelluca, 1987; Zebarth and others, 1998), including use in the San Joaquin Valley (Domagalski and Dubrovsky, 1991). 1,2,3-TCP has been detected in groundwater in areas where fumigants have been used (Zogorski and others, 2006). Consequently, 1,2,3-TCP was included in the fumigants category in this report, but actually represents a fumigant synthesis byproduct.

Fumigants, as a class, were detected at high relative-concentrations in 3.6 and 2.1 percent of the primary aquifers in the SESJ and the KERN study units, respectively (table 12). Four fumigants—DBCP, EDB, 1,2-DCP, and 1,2,3-TCP—were detected in USGS-grid wells in either the SESJ or the KERN study units (Burton and Belitz, 2008; Shelton and others, 2008). Only DBCP and 1,2-dibromoethane (EDB) were detected at high or at moderate relative-concentrations (table 10, fig. 23F). In general, fumigants were detected more frequently in the KERN study unit than in the SESJ study unit except for DBCP (fig. 23E); most of the high detections of DBCP in either USGS-grid wells or in CDPH-other wells were in the Kings study area (fig. 24B). DBCP also was detected at high relative-concentrations near the city of Delano and in two locations in the KERN study unit.

DBCP

Historically, DBCP was used as a soil fumigant to control nematodes. Between 1955 and 1977, DBCP primarily was used on orchards and on vineyards but also on some row crops in California, including the Fresno area (Peoples and others, 1980; Domagalski, 1997; Burow and others, 1999). Use of DBCP was discontinued by the California Department of Food and Agriculture in 1977 in response to concern about the potential hazardous effects of DBCP on human health (Domalgoski, 1997; Burow and others, 1999).

DBCP concentrations were positively correlated to lateral position and DO and negatively correlated to pH in the SESJ study unit (table 13). When the data from the Domestic Well Project was included, DBCP significantly correlated with orchard and vineyard land use (table 13). DBCP was not correlated to any of the explanatory factors in the KERN study unit. The absence of correlations of DBCP with explanatory variables may be a result of the low detection frequency for DBCP in the KERN study unit, making it difficult to identify relations.

DBCP concentration data from the Priority Basin Project was not correlated with orchard and vineyard land use without the data from the GAMA Domestic Well Project. This is in contrast to two Priority Basin Project study units in the San Joaquin Valley (Bennett and others, 2010; Landon and others, 2010a) which found significant positive correlations with orchard and vineyard land use in the Priority Basin Project data. However, the detection frequency of DBCP in the SESJ study-unit grid wells (33 percent) was significantly higher when the orchard and vineyard land use was greater than 40 percent than in the SESJ study-unit grid wells (15 percent) when orchard and vineyard land use was less than 40 percent (fig. 27); this indicates that orchard and vineyard land use may have some effect on the presence of DBCP.

The detection frequency of DBCP in wells where DO was greater than 2.0 mg/L was significantly higher than in wells where the DO was less than 2.0 mg/L (fig. 27). The higher detection frequency of DBCP with higher DO concentrations may partially be because DBCP is resistant to biological transformation in oxic conditions (Bloom and Alexander, 1990; Burow and others, 1999). Not only are DO and lateral position correlated with each other, but they are correlated to orchard and vineyard land use. This provides additional indirect evidence that orchard and vineyard land use may affect where DBCP may be detected. DBCP also was correlated with pH, but this relation may result from the correlation of pH with DO and orchard and vineyard land use (table 6B).

DBCP concentrations were not correlated with well depth, depth to top-of-perforations, or groundwater age. However, detection frequencies of DBCP in modern-age groundwater (31 percent) were greater than in pre-modern-age groundwater (10 percent). The presence of DBCP in older groundwater may indicate the ability of DBCP to persist in the aquifer as a result of the low organic content of the aquifer materials (Burow and others, 1999). Another explanation for the occurrence of DBCP in pre-modern water is the presence of short-circuit mechanisms as a result of well construction or well operation practices that allow modern contaminants to mix with deeper, pre-modern water, as has been found for other constituents (Jurgens and others, 2008; Landon and others, 2010b).

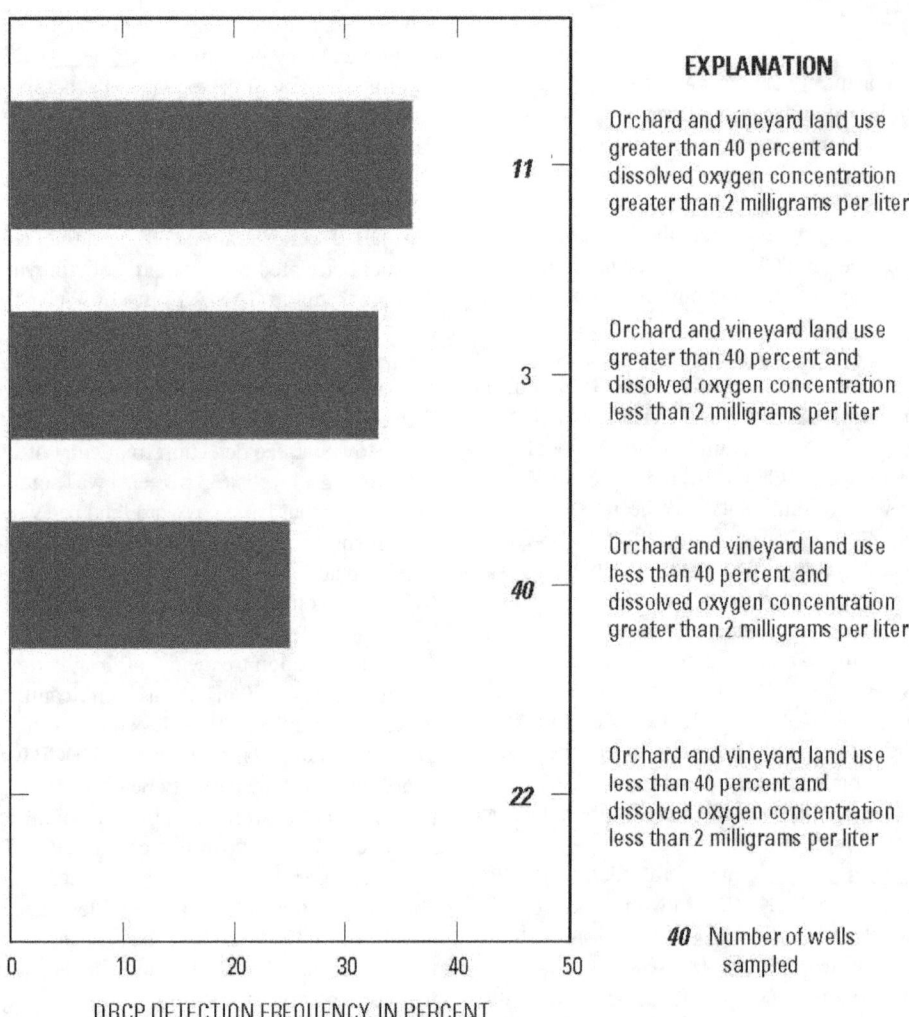

EXPLANATION

Orchard and vineyard land use greater than 40 percent and dissolved oxygen concentration greater than 2 milligrams per liter

Orchard and vineyard land use greater than 40 percent and dissolved oxygen concentration less than 2 milligrams per liter

Orchard and vineyard land use less than 40 percent and dissolved oxygen concentration greater than 2 milligrams per liter

Orchard and vineyard land use less than 40 percent and dissolved oxygen concentration less than 2 milligrams per liter

40   Number of wells sampled

DBCP DETECTION FREQUENCY, IN PERCENT

Figure 27.   Relation of DBCP detection frequency to percentage of orchard and(or) vineyard land use and dissolved oxygen, Southeast San Joaquin Valley study unit, California GAMA Priority Basin Project.

Pesticides

Pesticides are used in agricultural and urban settings. Pesticides, as a class, were not detected at high relative-concentrations in the study units but were detected at moderate relative-concentrations in 2.4 percent of the primary aquifers in the SESJ study unit (table 12). The herbicides atrazine and simizine were detected at low relative-concentrations in more than 10 percent of the primary aquifers in both study units (fig. 23C). Bromacil and dinoseb were detected in more than 10 percent of the primary aquifers in SESJ and KERN, respectively. Atrazine and simazine were the most commonly detected, with detection frequencies over 30 percent in the study units. Atrazine and simazine were among the most commonly detected herbicides in groundwater in major aquifers across the United States (Gilliom and others, 2006). Historically, simazine is most commonly used on vineyards and orchards in the study units, but also is used on rights-of-way for weed control (Domagalski and Dubrovsky, 1991). Dieldrin, an insecticide, was detected at moderate concentrations in more than 2 percent of the primary aquifers in both study units (table 9; figs. 23C and 23D). Dieldrin still persists in the environment even though its use was discontinued in California in 1987 (Barbash and Resek, 1996).

Similar to VOCs, more than one pesticide usually was detected in wells with pesticides (fig. 24C). The number of pesticides detected in a well was correlated with well depth and groundwater age in both study units (tables 5 and 13). The number of pesticides in a well was positively correlated with DO, percentage of urban land use, lateral position, and septic system density in the SESJ study unit (table 13). The number of pesticides was negatively correlated with depth to top-of-perforations, pH, and percentage of agricultural land use in the SESJ study unit (table 13).

The number of pesticides detected was significantly lower in pre-modern-age groundwater than in modern-age groundwater for the study units (fig. 28; table 5). However, pesticides were detected in 42 and 50 percent of the wells with pre-modern-age groundwater in the SESJ and the KERN study units, respectively. This indicates that some pesticides may persist in groundwater for long periods of time.

Pesticides in the SESJ study unit were detected more frequently in wells with depths to the top-of-perforations less than 250 ft (detection frequency 83 percent compared to 45 percent). However, pesticides were detected in some wells with depths to top-of-perforations as great as 1,000 ft (fig. 28A). Pesticides in the KERN study unit were detected more frequently in wells with depths to top-of-perforations less than 400 ft (detection frequency 86 percent compared to 63 percent). However, pesticides were detected in a few wells with depths to top-of-perforations as great as 640 ft (fig. 28B). The presence of pesticides in deep wells could indicate the effects of short-circuit mechanisms because of well construction or well operation practices that allow modern contaminants to mix with pre-modern water, as has been found for other constituents (Jurgens and others, 2008; Landon and others, 2010b)

The correlation of DO with pesticides in the SESJ study unit likely is a result of the correlation of DO with well depth (table 6B). The positive correlation of pesticide detections with lateral position indicates that the number of pesticides detected increases eastward through the study unit. A majority of the detections were in the eastern part of the SESJ study unit (fig. 24C).

The number of pesticides detected was positively correlated with urban land use in the SESJ study unit. Herbicide concentrations also were found to be positively correlated to urban land use in the Priority Basin Project Central Eastside study unit (Landon and others, 2010a). In contrast, the number of pesticides detected was negatively correlated with agricultural land use. This surprising correlation most likely is a result of the correlation of number of pesticides detected with depth to top-of-perforations. The wells in the SESJ study unit with agricultural land use are deeper than wells with urban land use. In the SESJ study unit, changes in land use from east to west across the study unit make it feasible that historical pesticide use patterns have not been uniform across the study unit. In addition, the Corcoran clay layer located in the western part of the SESJ study unit can act as a barrier to the downward migration of pesticides into the deeper aquifer used by many CDPH and irrigation wells.

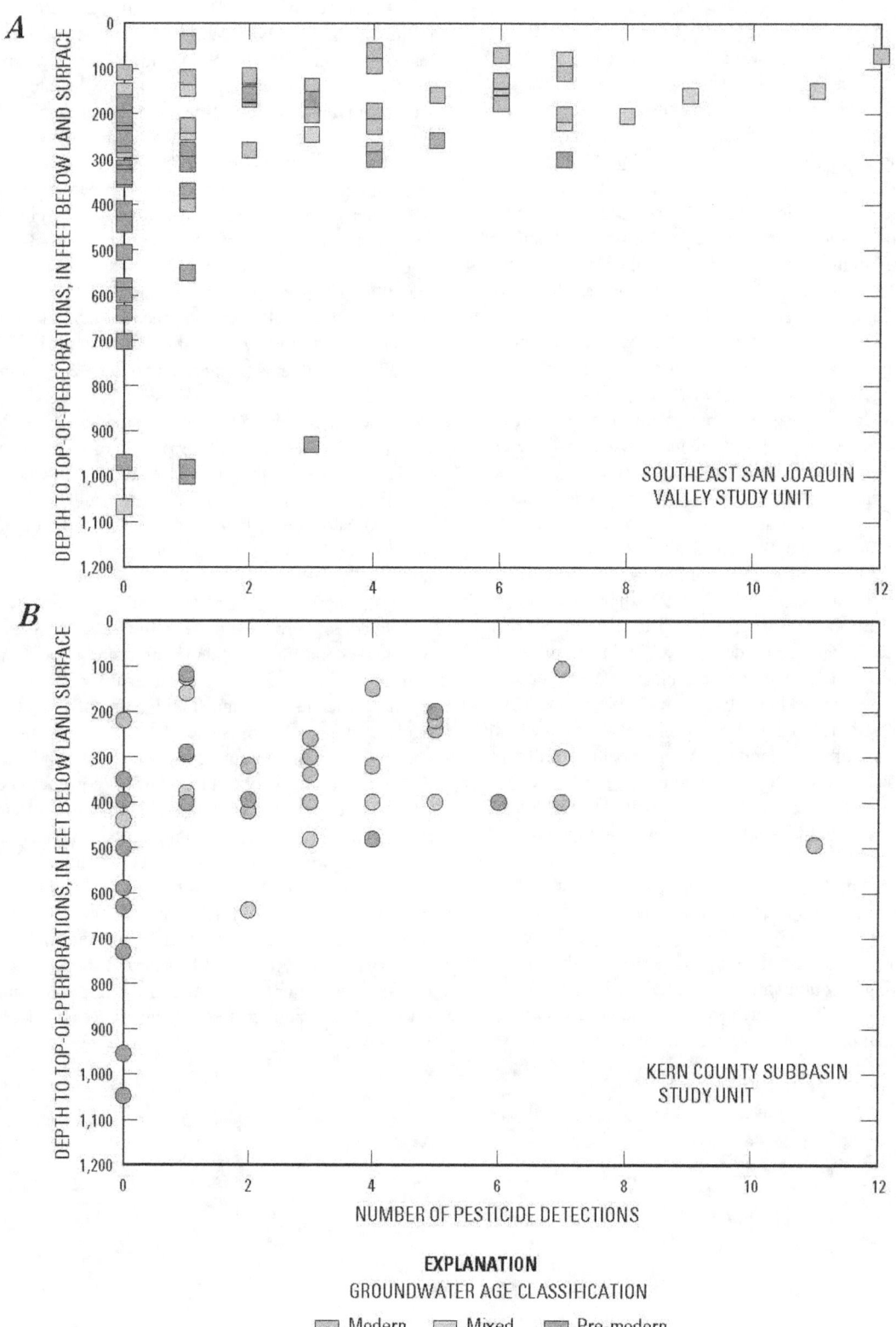

Figure 28. (A–B) Relation of number of pesticide detections to depth to top-of-perforations and groundwater age in the two southern San Joaquin Valley study units, California GAMA Priority Basin Project.

Constituents of Special Interest

Special-interest constituents, similar to organic constituents, usually are anthropogenic in origin. The special-interest constituents analyzed by the Priority Basin Project in the southern San Joaquin Valley are perchlorate and *N*-nitrosodimethylamine (NDMA). Possible anthropogenic sources of perchlorate could include nitrate fertilizers mined from the Atacama Desert of Chile that have been used historically on some orchard crops (Dasgupta and others, 2006), or industrial, manufacturing, or commercial uses such as explosives, road flares, rocket fuel, and other products (California Department of Public Health, 2008c; Parker and others, 2008). Perchlorate can occur under natural conditions in a variety of climatic conditions (Fram and Belitz, 2011), and not just arid climates (Dasgupta and others, 2005; Plummer and others, 2006). However, perchlorate is more likely to occur naturally in arid environments such as the arid and semi-arid environments found in the southwestern United States (Fram and Belitz, 2011). Perchlorate and NDMA have been detected recently in, or are considered to have the potential to reach, water resources used for drinking-water supplies (California Department of Public Health, 2008b, 2008c).

Constituents of special interest, as a class, were detected at high relative-concentrations in 1.2 percent of the primary aquifers in the SESJ study unit, and at moderate relative-concentrations in 19 and 6.4 percent in the SESJ and the KERN study units, respectively (table 12). Perchlorate was detected at high relative-concentrations (1.2 percent) in the SESJ study unit but not in the KERN study unit. Perchlorate was detected at moderate relative-concentrations in 18 and 6.4 percent in the SESJ and the KERN study units, respectively (table 9; figs. 23G and 23H). NDMA was detected at moderate relative-concentrations in 3.4 percent of the primary aquifers in the SESJ study unit (table 9; fig. 23H). Detection frequencies of perchlorate were almost 20 percent in the SESJ study unit with detections located in the Kings, Kaweah, and Tule study areas (figs. 23E and 24D).

Perchlorate

Perchlorate was positively correlated to DO in the study units (table 13). Perchlorate was not detected in any samples where the DO was less than 3.0 mg/L (fig. 29). Perchlorate also was positively correlated to density of LUFTs and septic systems around the wells and lateral position, and negatively correlated to well depth, depth to top-of-perforations, and pH in the SESJ study unit (table 13). Perchlorate concentrations were not significantly different between groundwater age categories (table 5).

Perchlorate concentrations decreased as well depth increased in the SESJ study unit. In contrast, perchlorate concentrations were not correlated with well depth in the Central Eastside study unit which also is the Priority Basin Project study unit in the San Joaquin Valley (Landon and others, 2010a). However, the positive correlations of perchlorate to DO in the study units were similar in the Central Eastside study unit. The positive correlation of perchlorate and DO could be affected by perchlorate biodegradation under anoxic conditions (Sturchio and others, 2007) or may simply indicate that concentrations for both constituents typically are high in shallow groundwater. The correlation with lateral position may be influenced by the correlation of lateral position with DO in the SESJ study unit. Perchlorate was not correlated to orchard and vineyard land use. This lack of correlation was not expected because Chilean fertilizer, a source of perchlorate, was used historically on orchards. The correlations of LUFTs and septic systems with perchlorate (table 6B) likely indicate unknown variations in sources of perchlorate across the study unit.

The predicted probability of detecting naturally occurring perchlorate at a concentration greater than 0.5 µg/L is 10 to 20 percent on the basis of the logistic regression model developed by Fram and Belitz (2011) in the SESJ and KERN study units. The predicted probability of detecting naturally occurring perchlorate at a concentration greater than 1 µg/L is only 1 to 5 percent. All the perchlorate detections were greater than 0.5 µg/L, and more than one-third of the perchlorate detections were greater than 1 µg/L. This indicates that anthropogenic sources have contributed perchlorate to groundwater in the study units.

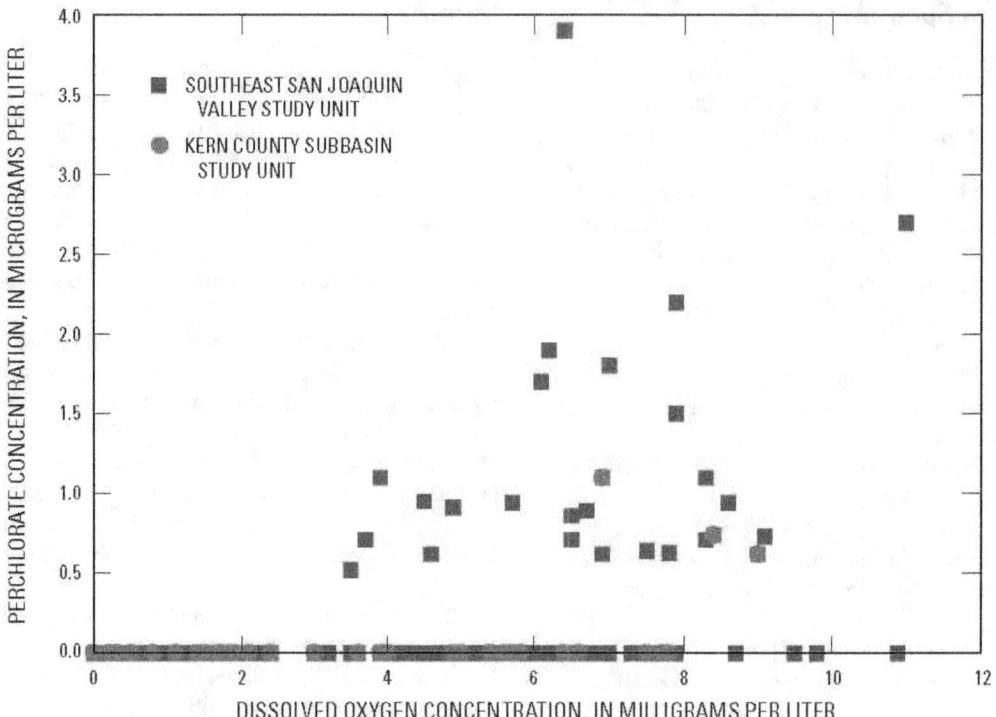

Figure 29.   Relation of perchlorate concentration to dissolved oxygen in the two southern San Joaquin Valley study units, California GAMA Priority Basin Project.

Summary

Groundwater quality in the southern San Joaquin Valley was investigated as part of the Priority Basin Project of the Groundwater Ambient Monitoring and Assessment (GAMA) Program. The GAMA Priority Basin Project is conducted by the U.S. Geological Survey (USGS) in collaboration with the California State Water Resources Control Board and the Lawrence Livermore National Laboratory. Two Priority Basin Project study units are located in the southern San Joaquin Valley—Southeast San Joaquin Valley (SESJ) and Kern County Subbasin (KERN).

The GAMA Priority Basin Project is designed to provide a statistically unbiased, spatially distributed assessment of untreated groundwater quality in the primary aquifers at the basin-scale. The aquifer systems (hereinafter referred to as primary aquifers) were defined as that part of the aquifer corresponding to the perforation interval of wells listed in the CDPH database for the study units. Wells were randomly selected within spatially distributed grid cells across the study units to assess the quality of the groundwater. Samples were collected by the USGS from October 2005 through March 2006 from 130 grid wells (83 in the SESJ study unit and 47 in the KERN study unit) which included 108 CDPH wells, 15 irrigation wells, 6 domestic wells, and 1 fire protection well. An additional 19 wells (8 CDPH wells and 11 monitoring wells) were sampled to improve the understanding of the relation of water quality to explanatory factors. Samples from USGS-grid and USGS-understanding wells were analyzed for up to 345 constituents. CDPH inorganic data from the 3-year period (January 1, 2003 to December 31, 2005) were used to complement USGS-grid well data and provide additional information about groundwater quality.

Relative-concentrations (sample concentration divided by water-quality benchmark concentration) were used for evaluating groundwater quality. Selected constituents with high relative-concentrations or detection frequencies greater than or equal to 10 percent were selected to focus the understanding assessment on those constituents that have the greatest effect on water quality. The relative-concentration threshold for classifying inorganic constituents as moderate was 0.5, whereas for organic constituents it was 0.1. A relative-concentration of 0.1 was used as a boundary between low and moderate values of organic and special-interest constituents for consistency with other studies and reporting requirements.

Aquifer-scale proportion was used as a metric for assessing the quality of untreated groundwater for the study units. High aquifer-scale proportion is defined as the areal percentage of the primary aquifers with a relative-concentration greater than 1.0. Moderate and low aquifer-scale proportions were defined as the areal percentage of the primary aquifers with moderate and low relative-concentrations, respectively. Grid-based and spatially weighted statistical approaches were used to assess aquifer-scale proportions of constituents at high, moderate, and low relative-concentrations in the primary aquifers.

Inorganic constituents were more prevalent and relative-concentrations for inorganic constituents generally were higher than for organic constituents. For inorganic constituents with human-health benchmarks, relative-concentrations for 30 and 23 percent of the primary aquifers were high for at least one constituent in the SESJ and the KERN study units, respectively. In the study units, the inorganic constituents with human-health benchmarks that were detected at high relative-concentrations in more than 2 percent of the primary aquifers were arsenic, boron, vanadium, nitrate, uranium, and unadjusted gross alpha radioactivity. Additional constituents with human-health benchmarks—antimony, radium, and fluoride—were detected at high relative-concentrations in the KERN study unit.

For inorganic constituents with aesthetic benchmarks (SMCL), relative-concentrations for 6.6 and 22 percent of the primary aquifers were high for at least one constituent in the SESJ and the KERN study units, respectively. Inorganic constituents with aesthetic benchmarks that were detected at high relative-concentrations in the study units were iron and manganese. Additional constituents with aesthetic benchmarks—total dissolved solids (TDS), sulfate, and chloride—were detected at high relative-concentrations in KERN.

In contrast to inorganic constituents, organic constituents with human-health benchmarks were detected at high relative-concentrations in 4.8 and 2.1 percent of the primary aquifers in the SESJ and KERN study units, respectively. Of the 78 VOCs analyzed (not including fumigants), 28 were detected—23 with human-health benchmarks. Benzene, carbon tetrachloride, tetrachloroethene (PCE), and trichloroethene (TCE) were the only VOCs detected at high relative-concentrations. Dichloromethane was detected at moderate relative-concentrations in the SESJ study unit, and 1,2-dichloroethane was detected at moderate relative-concentrations in the KERN study unit. Human-health benchmarks were established for all five fumigants detected. Fumigants were detected at high relative-concentrations in 3.6 percent and 2.1 percent of the SESJ and KERN study units, respectively. The fumigant, 1,2-dibromo-3-chloropropane (DBCP), was detected at high relative-concentrations in both study units. Of the 136 pesticides and polar pesticides analyzed, 33 pesticides were detected—18 with and 15 without human-health benchmarks. All pesticides were detected at low relative-concentrations or were not detected except dieldrin. Dieldrin was detected at moderate relative-concentrations in both study units. The detection frequencies for two pesticides, simazine and atrazine, were greater than or equal to 10 percent in the

study units. The special-interest constituent, perchlorate, was detected at high and moderate relative-concentrations in the SESJ study unit and at moderate relative-concentrations in the KERN study unit.

The understanding assessment used statistical correlations between concentrations of constituents and values of selected potential explanatory factors to identify the factors potentially affecting the concentrations and occurrences of inorganic or organic constituents detected at high relative-concentrations or for organic constituents with detection frequencies greater than 10 percent. The potential explanatory factors evaluated were land use, depth, lateral position in flow path, septic system density, formerly leaking underground fuel tanks (LUFTs), groundwater age, and geochemical conditions [dissolved oxygen (DO) and pH]. The datasets from the two study units were treated separately.

Well depth, depth to top-of-perforations, groundwater age, DO, and pH were the explanatory factors that most frequently were correlated with inorganic constituents. Depth, groundwater age, land use, and DO were the explanatory factors that most frequently were correlated to organic and special-interest constituents.

Arsenic concentrations were significantly high in deep and old (pre-modern) groundwater in the SESJ study unit, but not in the KERN study unit. Arsenic also was correlated with geochemical conditions. Arsenic concentrations tended to increase when pH was above 7.6 or when DO decreased in the SESJ study unit. Antimony concentrations also increased with pH in the SESJ study unit. Similar to arsenic, boron concentrations were significantly higher in old, deep groundwater in the SESJ study unit. Boron concentrations also increased as pH increased in the SESJ study unit. Vanadium concentrations also were positively correlated to DO in the SESJ study unit, but were not correlated to depth or groundwater age. Similarly, fluoride concentrations generally increased in deep or old groundwater, with high pH, and with low DO in the study units. Manganese and iron concentrations, like the trace elements, increased with increasing well depth and decreasing DO. In contrast to most of the other trace elements, uranium concentrations were high in shallow wells with modern-age groundwater and low pH. The source of all these constituents are attributed to natural sources. However, elevated uranium concentrations in shallow groundwater was attributed to the enhanced desorption of uranium from sediments by irrigation and urban recharge with high bicarbonate concentrations.

Nitrate concentrations, in contrast to most of the trace elements, were high in shallow wells with modern-age groundwater. Nitrate concentrations increased with increasing DO. Nitrate was not correlated with agricultural land use as in some other studies, but nitrate was positively correlated with orchard and vineyard land use (a subset of agricultural land use). Nitrate occurs naturally in the study units, but concentrations are elevated by human activities.

In general, concentrations of TDS, sulfate, and chloride were significantly higher in the KERN study unit than in the SESJ study unit. TDS was negatively correlated with pH in the study units. TDS and sulfate were positively correlated to agricultural land use in the KERN study unit, but not in the SESJ study unit. Sulfate concentrations were high in shallow wells in the primary aquifers in the SESJ study unit whereas chloride concentrations were high in deep wells in the KERN study unit. These correlations suggest TDS and sulfate concentrations in SESJ primarily are influenced by human factors while TDS and sulfate concentration in KERN study unit primarily are influenced by natural sources.

Organic compounds usually were detected at low relative-concentrations; therefore, statistical analyses of relations to explanatory factors usually were done for classes of constituents. Classes of organic compounds consisted of VOCs—which were further subdivided into trihalomethanes (THMs), solvents, and other VOCs—fumigants, and pesticides. VOCs were detected more frequently in shallow wells with modern-age groundwater and high DO in the SESJ study unit. These relations were not observed in the KERN study unit.

Total THM concentration and solvent detection frequency decreased in deep and old (pre-modern) groundwater in the SESJ study unit. Statistical relations with depth were absent in the KERN study unit possibly because relatively few shallow wells were sampled in the KERN study unit. THM concentrations and solvent detection frequency were positively correlated with DO in the SESJ study unit. Solvent detection frequency in the SESJ study unit was positively correlated with urban land use, whereas THM concentration was not correlated with urban land use as has been observed in other Priority Basin Project study units.

High relative-concentrations of DBCP accounted for the large majority of fumigant detections; therefore, statistical analysis was performed on DBCP concentrations to represent fumigants as a class. DBCP was positively correlated to DO in the SESJ study unit. DBCP was not correlated to orchard and vineyard land use as observed in two other Priority Basin Project study units; however, the detection frequency of DBCP was significantly higher when the orchard and vineyard land use was greater than 40 percent compared to when orchard and vineyard land use was less than 40 percent. DBCP, unlike many other organic compounds, was not correlated with depth.

Pesticide detection frequency, similar to other organic constituents, decreased in deep or old (pre-modern) groundwater in the SESJ study unit. Pesticide detections also decreased with increasing well depth in the KERN study unit. Pesticide detections in the SESJ study unit were positively correlated with urban land use and negative correlated with agricultural land use.

Similar to organic constituents, perchlorate concentrations decreased with increasing well depth in the SESJ study unit but were not correlated to groundwater age. Perchlorate concentrations generally were high in groundwater with high DO conditions. Anthropogenic sources have contributed perchlorate to groundwater in the study units although low levels of perchlorate may occur naturally.

There are many similarities and differences between the SESJ and the KERN study units. In the study units, relative-concentrations of arsenic, vanadium, boron, uranium, gross alpha radioactivity, nitrate, manganese, and iron were high in more than 2 percent of the primary aquifers. However, vanadium concentration was significantly higher in the SESJ study unit than in the KERN study unit. In addition, in the KERN study unit, relative-concentrations of antimony, fluoride, TDS, sulfate, and chloride were high in more than 2 percent of the primary aquifers. In the KERN study unit, concentrations of TDS, sulfate, and chloride were significantly higher than in the SESJ study unit, although concentrations of TDS and chloride in the SESJ study unit were high in less than 1 percent of the primary aquifers.

The organic constituents, carbon tetrachloride, TCE, and DBCP, were detected at high relative-concentrations in the study units. However, DBCP concentrations were significantly higher in the SESJ study unit than in the KERN study unit. In addition, benzene, PCE, 1,2-dibromoethane (EDB), and perchlorate were detected at high relative-concentrations in the SESJ study unit but not in the KERN study unit. Although simazine and atrazine were detected at low relative-concentrations in more than 10 percent of the primary aquifers in the study units, bromacil was detected in more than 10 percent of the primary aquifers in the SESJ study unit, and dinoseb was detected in more than 10 percent of the primary aquifers in the KERN study unit.

Some of the explanatory factors were significantly different between the two study units. Well depth and depth to top-of-perforations were significantly deeper in the KERN study unit than in the SESJ study unit. In contrast, DO and percentage of urban land use were significantly higher in the SESJ study unit than in the KERN study unit. pH was not significantly different between the KERN study unit and the SESJ study unit, but the pH values in the KERN study unit are skewed toward higher pH values than in the SESJ study unit. These differences in explanatory factors may explain some of the dissimilarity in relations between water-quality variables and explanatory factors observed between the two study units.

VOCs, pesticides, and perchlorate may be used as tracers of groundwater that has recharged over the decades because these compounds began to be used for industrial and commercial purposes because of the low concentration at which these compounds were detected. Low-level analyses provide an early awareness of constituents whose presence in groundwater at low concentrations may be important for the prioritization of monitoring water quality in the future.

Acknowledgments

The authors thank the following cooperators for their support: the California State Water Resources Control Board, Lawrence Livermore National Laboratory, California Department of Public Health, and California Department of Water Resources. We especially thank the well owners and water purveyors for their cooperation and generosity in allowing the USGS to collect samples from their wells. Funding for this work was provided by State of California bonds authorized by Proposition 50 and administered by the State Water Resources Control Board.

References Cited

Aeschbach-Hertig, W., Peeters, F., Beyerle, U., and Kipfer, R., 1999, Interpretation of dissolved atmospheric noble gases in natural waters: Water Resources Research, v. 35, no. 9, p. 2779–2792.

Aeschbach-Hertig, W., Peeters, F., Beyerle, U., and Kipfer, R., 2000, Paleotemperature reconstruction from noble gases in ground water taking into account equilibration with entrapped air: Nature, v. 405, p. 1040–1044.

Agency for Toxic Substances and Disease Registry, 1995, ToxFAQs for antimony and compounds: Agency for Toxic Substances and Disease Registry website, accessed March 14, 2011, at http://www.atsdr.cdc.gov/toxfaqs/tf.asp?id=331&tid=58.

Andrews, J.N., 1985, The isotopic composition of radiogenic helium and its use to study groundwater movement in confined aquifers: Chemical Geology, v. 49, p. 339–351.

Andrews, J.N., and Lee, D.J., 1979, Inert gases in groundwater from the Bunter Sandstone of England as indicators of age and paleoclimatic trends: Journal of Hydrology, v. 41, p. 233–252.

Barbash, J.E., and Resek, E.A., 1996, Pesticides in ground water—Distribution, trends, and governing factors: Pesticides in the Hydrologic Systems, v. 2, 588 p.

Belitz, K., Dubrovsky, N.M., Burow, K.R., Jurgens, B., and Johnson, T., 2003, Framework for a ground-water quality monitoring and assessment program for California: U.S. Geological Survey Water-Resources Investigations Report 03-4166, 78 p. (Also available at http://pubs.usgs.gov/wri/wri034166/.)

Belitz, K., Jurgens, B., Landon, M.K., Fram, M.S., and Johnson, T., 2010, Estimation of aquifer-scale proportion using equal-area grids—Assessment of regional-scale groundwater quality: Water Resources Research, v. 46, citation number W11550, doi:10.1029/2010WR009321, accessed March 14, 2011, at http://www.agu.org/pubs/crossref/2010/2010WR009321.shtml.

Bennett, G.L., Fram, M.S., Belitz, K., and Jurgens, B.C., 2010, Status and understanding of groundwater quality in the northern San Joaquin basin, 2005—California GAMA Priority Basin Project: U.S. Geological Survey Scientific Investigations Report 2010-5175, 82 p. (Also available at http://pubs.usgs.gov/sir/2010/5175/.)

Bloom, R.A., and Alexander, M., 1990, Microbial transformation of 1,2-dibromo-3-chloropropane (DBCP): Journal of Environmental Quality, v. 19, p. 722–726.

Brown, L.D., Cai, T.T., and Dasgupta, A., 2001, Interval estimation for a binomial proportion: Statistical Science, v. 16, no. 2, p. 101–133, doi:10.1214/ss/1009213286, accessed March 14, 2011, at http://projecteuclid.org/euclid.ss/1009213286.

Burow, K.R., Dubrovsky, N.M., and Shelton, J.L., 2007, Temporal trends in concentrations of DBCP and nitrate in groundwater in the eastern San Joaquin Valley, California, USA: Hydrogeology Journal, v. 15, no. 5, p. 991–1007, doi:10.1007/s10040-006-0148-7, accessed March 14, 2011, at http://www.springerlink.com/content/wk64558523hr1833/.

Burow, K.R., Jurgens, B.C., Kauffman, L.J., Phillips, S.P., Dalgish, B.A., and Shelton, J.L., 2008, Simulations of ground-water flow and particle pathline analysis in the zone of contribution of a public-supply well in Modesto, eastern San Joaquin Valley, California: U.S. Geological Survey Scientific Investigations Report 2008-5035, 41 p. (Also available at http://pubs.usgs.gov/sir/2008/5035.)

Burow, K.R., Panshin, S.Y., Dubrovsky, N.M., VanBrocklin, D., and Fogg, G.E., 1999, Evaluation of processes affecting 1,2-dibromo-3-chloropropane (DBCP) concentrations in ground water in the eastern San Joaquin Valley, California—Analysis of chemical data and ground-water flow and transport simulations: U.S. Geological Survey Water-Resources Investigations Report 99-4059, 57 p. (Also available at http://ca.water.usgs.gov/sanj/pub/usgs/wrir99-4059/wrir99-4059.pdf.)

Burow, K.R., Shelton, J.L., and Dubrovsky, N.M., 1998, Occurrence of nitrate and pesticides in ground water beneath three agricultural land-use settings in the eastern San Joaquin Valley, California, 1993–1995: U.S. Geological Survey Water-Resources Investigations Report 97-4284, 51 p. (Also available at http://ca.water.usgs.gov/sanj/pub/usgs/wrir97-4284/wrir97-4284.pdf.)

Burow, K.R., Shelton, J.L., and Dubrovsky, N.M., 2008, Regional nitrate and pesticide trends in ground water in the eastern San Joaquin Valley, California: Journal of Environmental Quality, v. 37, no. 5, Supplement, p. S-249-S-263, doi:10.2134/jeq2007.0061, accessed March 14, 2011, at https://www.agronomy.org/publications/jeq/abstracts/37/5_Supplement/S-249.

Burow, K.R., Stork, S.V., and Dubrovsky, N.M., 1998, Nitrate and pesticides in ground water in the eastern San Joaquin Valley, California—Occurrence and trends: U.S. Geological Survey Water-Resources Investigations Report 98-4040A, 33 p.

Burow, K.R., Weissmann, G.S., Miller, R.D., and Placzek, G., 1997, Hydrogeologic facies characterization of an alluvial fan near Fresno, California, using geophysical techniques: U.S. Geological Survey Open-File Report 97-46, 15 p. (Also available at http://pubs.er.usgs.gov/djvu/OFR/1997/ofr_97_46.djvu.)

Burton, C.A., and Belitz, K., 2008, Ground-water quality data in the Southeast San Joaquin Valley, 2005–2006—Results from the California GAMA Program: U.S. Geological Survey Data Series 351, 103 p. (Also available at http://www.swrcb.ca.gov/gama/docs/se_sanjoaquin_dsr351.pdf.)

Burton, J.D., 1996, The ocean, a global geochemical system, in Summerhayes, C.P., and Thorpe, S.A., eds., Oceanography, an illustrated guide: New York, Wiley and Sons, p. 165–181.

California Department of Health Services, 2007, California Code of Regulations, Title 22, Social Security, Division 4, Environmental Health, Chapter 15, Domestic Water Quality and Monitoring Regulations: Register 2007, No. 4, accessed February 10, 2007, at http://ccr.oal.ca.gov/.

California Department of Public Health, 2008a, California drinking water-related laws: California Department of Public Health website, accessed June 1, 2008, at http://www.cdph.ca.gov/certlic/drinkingwater/Pages/Lawbook.aspx.

California Department of Public Health, 2008b, Perchlorate in drinking water: California Department of Public Health website, accessed January 2, 2008, at http://www.cdph.ca.gov/certlic/drinkingwater/Pages/Perchlorate.aspx.

California Department of Public Health, 2008c, NDMA and other Nitrosamines—Drinking water issues: California Department of Public Health website, accessed July 7, 2008, at http://www.cdph.ca.gov/certlic/drinkingwater/Pages/NDMA.aspx.

California Department of Public Health, 2012, California Code of Regulation, Title 22, Division 4 Environmental Health, Chapter 15 Domestic water quality and monitoring regulations, accessed February 28, 2012 at http://ccr.oal.ca.gov.

California Department of Water Resources, 2003, California's groundwater: California Department of Water Resources Bulletin 118, 246 p, accessed March 23, 2011, at http://www.water.ca.gov/groundwater/bulletin118/bulletin118update2003.cfm.

California Department of Water Resources, 2006a, San Joaquin Valley Groundwater Basin—Kings Subbasin: California Department of Water Resources Bulletin 118, accessed March 23, 2011, at http://www.water.ca.gov/pubs/groundwater/bulletin_118/basindescriptions/5-22.08.pdf.

California Department of Water Resources, 2006b, San Joaquin Valley Groundwater Basin—Kaweah Subbasin: California Department of Water Resources Bulletin 118, accessed March 23, 2011, at http://www.water.ca.gov/pubs/groundwater/bulletin_118/basindescriptions/5-22.11.pdf.

California Department of Water Resources, 2006c, San Joaquin Valley Groundwater Basin—Tule Subbasin: California Department of Water Resources Bulletin 118, accessed March 23, 2011, at http://www.water.ca.gov/pubs/groundwater/bulletin_118/basindescriptions/5-22.13.pdf.

California Department of Water Resources, 2006d, San Joaquin Valley Groundwater Basin—Tulare Lake Subbasin: California Department of Water Resources Bulletin 118, accessed March 28, 2011, at http://www.water.ca.gov/pubs/groundwater/bulletin_118/basindescriptions/5-22.12.pdf.

California Department of Water Resources, 2006e, Cummings Valley Groundwater Basin: California Department of Water Resources Bulletin 118, accessed March 28, 2011, at http://www.water.ca.gov/pubs/groundwater/bulletin_118/basindescriptions/5-27.pdf.

California Department of Water Resources, 2008, Lines of equal elevation of water in wells, unconfined aquifer, San Joaquin Valley, Spring 2006: California Department of Water Resources website, accessed March 26, 2009, at http://www.water.ca.gov/pubs/groundwater/lines_of_equal_elevation_of_water_in_wells_unconfined_aquifer_san_joaquin_valley_spring_2008/sjv2008spr_unc_elev_color.pdf.

California Environmental Protection Agency, 2001, Geographic Environmental Information Management System GeoTracker (GEIMS) Leaking Underground Fuel/Storage Tank database (LUFT) [digital data]: Sacramento, Calif., California Environmental Protection Agency, State Water Resources Control Board, Division of Water Quality.

California Irrigation Management Information System, 2006, Daily temperature data for Bakersfield and Shafter stations, California, for period of record 1951–2006, accessed March 1, 2007, at http://www.cdph.ca.gov/certlic/drinkingwater/Pages/default.aspx.

California State Water Resources Control Board, 2003, A Comprehensive Groundwater Quality Monitoring Program for California: Report to the Governor and Legislature, March 2003, 121 p., accessed March 28, 2011, at http://www.waterboards.ca.gov/gama/docs/final_ab_599_rpt_to_legis_7_31_03.pdf.

California State Water Resources Control Board, 2011, GAMA—Groundwater Ambient Monitoring and Assessment Program: State Water Resources Control Board website, accessed March 14, 2011, at http://www.waterboards.ca.gov/water_issues/programs/gama.

Carlin, J.F., 2006, Antimony recycling in the United States in 2000: U.S Geological Survey Circular 1196-Q, 13 p., accessed March 28, 2011, at http://pubs.usgs.gov/circ/c1196q/.

Centers for Disease Control and Prevention, 2001, Recommendations for using fluoride to prevent and control dental caries in the United States: Morbidity and Mortality Weekly Report website, accessed March 14, 2011, at http://www.cdc.gov/mmwr/preview/mmwrhtml/rr5014a1.htm.

Cey, B.D., Hudson, G.B., Moran, J.E., and Scanlon, B.R., 2008, Impact of artificial recharge on dissolved noble gases in groundwater in California: Environmental Science and Technology, v. 42, p. 1017–1023.

Chapelle, F.H., 2001, Ground-water microbiology and geochemistry: New York, John Wiley and Sons, Inc., 477 p.

Chapelle, F.H., McMahon, P.B., Dubrovsky, N.M., Fuji, R.F., Oaksford, E.T., and Vroblesky, D.A., 1995, Deducing the distribution of terminal electron-accepting processes in hydrologically diverse groundwater systems: Water Resources Research, v. 31, no. 2, p. 359–371.

Clark, I.D., and Fritz, P., 1997, Environmental isotopes in hydrogeology: New York, Lewis Publishers, 328 p.

Craig, H., and Lal, D., 1961, The production rate of natural tritium: Tellus, v. 13, p. 85–105.

Dale, R.H., French, J.J., and Gordon, G.G., 1966, Ground-water geology and hydrology of the Kern River alluvial-fan area, California: U.S. Geological Survey, Open-File Report 66-21, 92 p.

Dasgupta, P.K., Dyke, J.V., Kirk, A.B., and Jackson, W.A., 2006, Perchlorate in the United States—Analysis of relative source contributions to the food chain: Environmental Science and Technology, v. 40, p. 6608–6614, doi: 10.1021/es061321z, accessed March 14, 2011, at http://pubs.acs.org/doi/abs/10.1021/es061321z.

Dasgupta, P.K., Martinelango, P.K., Jackson, W.A., Anderson, T.A., Tian, K., Tock, R.W., and Rajagopalan, S., 2005, The origin of naturally occurring perchlorate—The role of atmospheric processes: Environmental Science and Technology, v. 39, p. 1569–1575.

Davis, G.H., Green, J.H., Olmsted, F.H., and Brown, D.W., 1959, Groundwater conditions and storage capacity in the San Joaquin Valley, California: U.S. Geological Survey Water-Supply Paper 1469, 277 p.

Davis, S., and DeWiest, R.J., 1966, Hydrogeology: John Wiley and Sons, New York, 413 p.

Domagalski, J.L., 1997, Pesticides in surface and ground water of the San Joaquin-Tulare basins, California; analysis of available data, 1966 through 1992: U.S. Geological Survey Water-Supply Paper 2468, 24 p.

Domagalski, J.L., and Dubrovski, N.M., 1991, Regional assessment of non-point source pesticide residues in ground water, San Joaquin Valley, California: U.S. Geological Survey Water-Resources Investigations Report 91-4027, 64 p.

Dubrovsky, N.M., Kratzer, C.R., Brown, L.R., Gronberg, J.M., and Burow, K.R., 1998, Water quality in the San Joaquin-Tulare Basins, California, 1992–95: U.S. Geological Survey Circular 119, 38 p.

Duce, R.A., and Hoffman, G.L., 1976, Atmospheric vanadium transport to the ocean: Atmospheric Environment, v. 10, no. 11, p. 989–996.

Faunt, C.C., ed., 2009, Groundwater availability of the central valley aquifer, California: U.S. Geological Survey Professional Paper 1766, 225 p. (Also available at http://pubs.usgs.gov/pp/1766/.)

Fontes, J.C., and Garnier, J.M., 1979, Determination of the initial ^{14}C activity of the total dissolved carbon—A review of the existing models and a new approach: Water Resources Research, v. 15, p. 399–413.

Fram, M.S., and Belitz, K., 2011, Probability of detecting perchlorate under natural conditions in deep groundwater in California and the southwestern United States: Environmental Science and Technology, v. 45, p. 1271–1277.

Gilliom, R.J., Barbash, J.E., Crawford, C.G., Hamilton, P.A., Martin, J.D., Nakagaki, N., Nowell, L.H., Scott, J.C., Stackelberg, P.E., Thelin, G.P., and Wolock, D.M., 2006, The quality of our Nation's waters—Pesticides in the Nation's streams and ground water, 1992–2001: U.S. Geological Survey Circular 1291, 172 p. (Also available at http://pubs.usgs.gov/circ/2005/1291/.)

Helsel, D.R., and Hirsch, R.M., 2002, Statistical methods in Water Resources: U.S. Geological Survey Techniques of Water-Resources Investigations, Book 4, Chapter A3, 510 p. (Also available at: http://water.usgs.gov/pubs/twri/twri4a3/.)

Hem, J.D., 1970, Study and interpretation of the chemical characteristics of natural water (2d ed.): U.S. Geological Survey Water-Supply Paper 1473, 363 p.

Hem, J.D., 1985, Study and interpretation of the chemical characteristics of natural water (3d ed.): U.S. Geological Survey Water-Supply Paper 2254, 263 p.

Hope, B.K., 1997, An assessment of the global impact of anthropogenic vanadium: Biogeochemistry, v. 37, no. 1, p. 1–13.

Hutson, S.S., Barber, N.L., Kenny, J.F., Linsey, K.S., Lumia, D.S., and Maupin, M.A., 2004, Estimated use of water in the United States in 2000: U.S. Geological Survey Circular 1268, 46 p. (Also available at http://pubs.usgs.gov/circ/2004/circ1268/index.html.)

International Programme on Chemical Safety, 1988, Environmental Health Criteria 81, Vanadium, accessed March 16, 2011, at http://www.inchem.org/documents/ehc/ehc/ehc81.htm.

Isaaks, E.H., and Srivastava, R.M., 1989, Applied Geostatistics: New York, Oxford University Press, 561 p.

Izbicki, J.A., Stamos, C.L., Metzger, L.F., Halford, K.J., Kulp, T.R., and Bennett, G.L., 2008, Source, distribution, and management of arsenic in water from wells, eastern San Joaquin ground-water subbasin, California: U.S. Geological Survey Open-File Report 2008–1272, 8 p. (Also available at http://pubs.usgs.gov/of/2008/1272/.)

Jennings, C.W., 1977, Geologic map of California: California Department of Conservation, Division of Mines and Geology, Geologic Data Map, No. 2, scale 1:750,000.

Johnson, T.D., and Belitz, K., 2009, Assigning land use to supply wells for the statistical characterization of regional groundwater quality—Correlating urban land use and VOC occurrence: Journal of Hydrology, v. 370, p. 100–108.

Jurgens, B.C., Burow, K.R., Dalgish, B.A., and Shelton, J.L., 2008, Hydrogeology, water chemistry, and factors affecting the transport of contaminants in the zone of contribution to a public-supply well in Modesto, eastern San Joaquin Valley, California: U.S. Geological Survey Scientific Investigations Report 2008–5156, 78 p. (Also available at http://pubs.usgs.gov/sir/2008/5156/.)

Jurgens, B.C., Fram, M.S., Belitz, K., Burow, K.R., and Landon, M.K., 2010, Effects of ground-water development on uranium: Central Valley, California, USA: Ground Water, v. 48, no. 6, p. 913–928.

Jurgens, B.C., McMahon, P.B., Chapelle, F.H., and Eberts, S.M., 2009, An Excel® workbook for identifying redox processes in ground water: U.S. Geological Survey Open-File Report 2009–1004, 8 p. (Also available at http://pubs.usgs.gov/of/2009/1004/.)

Kalin, R.M., 2000, Radiocarbon dating of groundwater systems, in Cook, P.G., and Herczeg, A., eds., Environmental tracers in subsurface hydrology: Boston, Kluwer Academic Publishers, p. 111–144.

Kulongoski, J., and Belitz, K., 2004, Ground-water ambient monitoring and assessment program: U.S. Geological Survey Fact Sheet 2004-3088, 2 p.

Kulongoski, J.T., Hilton, D.R., Cresswell, R.G., Hostetler, S., and Jacobson, G., 2008, Helium-4 characteristics of groundwaters from Central Australia—Comparative chronology with chlorine-36 and carbon-14 dating techniques: Journal of Hydrology, v. 348, p. 176–194.

Landon, M.K., Belitz, K., Jurgens, B.C., Kulongoski, J.T., and Johnson, T.D., 2010, Status and understanding of groundwater quality in the Central Eastside San Joaquin Basin, 2006—California GAMA Program Priority Basin Project: U.S. Geological Survey Scientific Investigations Report 2009-5266, 97 p. (Also available at http://pubs.usgs.gov/sir/2009/5266/.)

Landon, M.K., Jurgens, B.C., Katz, B.G., Eberts, S.M., Burow, K.R., and Crandall, C.A., 2010, Depth-dependent sampling to identify short-circuit pathways to public-supply wells in multiple aquifer settings in the United States: Hydrogeology Journal, v. 18, no. 3, p. 577–593.

Lindburg, R.D., and Runnells, D.D., 1984, Groundwater redox reactions: Science, v. 225, p. 925–927.

Lofgren, B.E., and Klausing, R.L., 1969, Land subsidence due to ground-water withdrawal, Tulare-Wasco area, California: U.S. Geological Survey Professional Paper 437-B, 103 p.

Lucas, L.L., and Unterweger, M.P., 2000, Comprehensive review and critical evaluation of the half-life of tritium: Journal of Research of the National Institute of Standards and Technology, v. 105, no. 4, p. 541–549.

Manning, A.H., Solomon, D.K., and Thiros, S.A., 2005, $^3H/^3He$ age data in assessing the susceptibility of wells to contamination: Ground Water, v. 43, no. 3, p. 353–367.

McKelvey, V.E., Strobell, J.D., Jr., and Slaughter, A.L., 1986, The vanadiferous zone of the phosphoria formation in western Wyoming and southeastern Idaho (USA): U.S. Geological Survey Professional Paper 1465, 27 p.

McMahon, P.B., and Chapelle, F.H., 2008, Redox processes and water quality of selected principal aquifer systems: Ground Water, v. 46, no. 2, p. 259–271.

Michel, R.L., 1989, Tritium deposition in the continental United States, 1953–83: U.S. Geological Survey Water-Resources Investigations Report 89-4072, 46 p.

Michel, R.L., and Schroeder, R., 1994, Use of long-term tritium records from the Colorado River to determine timescales for hydrologic processes associated with irrigation in the Imperial Valley, California: Applied Geochemistry, v. 9, p. 387–401.

Moran, M.J., Zogorski, J.S., and Squillace, P.J., 2007, Chlorinated solvents in groundwater of the United States: Environmental Science and Technology, v. 41, no. 1, p. 74–81.

Morrison, P., and Pine, J., 1955, Radiogenic origin of the helium isotopes in rock: Annual New York Academy of Sciences, v. 12, p. 19–92.

Nakagaki, Naomi, Price, C.V., Falcone, J.A., Hitt, K.J., and Ruddy, B.C., 2007, Enhanced National Land Cover Data 1992 (NLCDe 92): U.S. Geological Survey Raster digital data, accessed March 28, 2011, at http://water.usgs.gov/lookup/getspatial?nlcde92.

Nakagaki, Naomi, and Wolock, D.M., 2005, Estimation of agricultural pesticide use in drainage basins using land cover maps and county pesticide data: U.S. Geological Survey Open-File Report 2005–1188, 46 p. (Also available at http://pubs.usgs.gov/of/2005/1188/.)

Nolan, B.T., and Hitt, K.J., 2006, Vulnerability of shallow groundwater and drinking-water wells to nitrate in the United States: Environmental Science and Technology, v. 40, no. 24, p. 7834–7840.

Nriagu, J.O., 1998, History, occurrence, and use of vanadium, in Nriagu, J.O., ed., Vanadium in the environment, Part 1—Chemistry and Biochemistry: New York, John Wiley and Sons, p. 1–24.

Oki, D.S., and Giambelluca, T.W., 1987, DBCP, EDB, and TCP contamination of groundwater in Hawaii: Groundwater, v. 25, no. 6, p. 193–702.

Page, R.W., 1973, Base of fresh ground water (approximately 3,000 micromhos) in the San Joaquin Valley, California: U.S. Geological Survey Hydrologic Investigations Atlas HA-489, 1 sheet, scale 1:500,000.

Page, R.W., 1986, Geology of the fresh ground-water basin of the Central Valley, California, with texture maps and sections: U.S. Geological Survey Professional Paper 1401-C, 54 p., 5 pl.

Parker, D.R., Seyfferth, A.L., and Kiel, B.K., 2008, Perchlorate in groundwater—A synoptic survey of "pristine" sites in the coterminous United States: Environmental Science and Technology, v. 42, no. 5, p. 1465–1471.

Pavelic, P., Dillon, P.J., and Nicholson, B.C., 2006, Comparative evaluation of the fate of disinfection byproducts at eight aquifer storage and recovery sites: Environmental Science and Technology, v. 40, no. 2, p. 501–508.

Peoples, S.A., Maddy, K.T., Cusick, W., Jackson, T., Cooper, C., and Frederickson, A.S., 1980, A study of samples of well water collected from selected areas in California to determine the presence of DBCP and certain other pesticide residues: Bulletin of Environmental Contamination and Toxicology, v. 24, p. 611–618.

Piper, A.M., 1944, A graphic procedure in the geochemical interpretation of water analyses: American Geophysical Union Transactions, v. 25, p. 914–923.

Plummer, L.N., Böhlke, J.K., and Doughten, M.W., 2006, Perchlorate in Pleistocene and Holocene groundwater in North-Central New Mexico: Environmental Science and Technology, v. 40, no. 6, p. 1757–1763.

Plummer, L.N., Michel, R.L., Thurman, E.M., and Glynn, P.D., 1993, Environmental tracers for age-dating young ground water, in Alley, W.M., ed., Regional ground-water quality: New York, Van Nostrand Reinhold, chap. 11, p. 255–294.

Poreda, R.J., Cerling, T.E., and Salomon, D.K., 1988, Tritium and helium isotopes as hydrologic tracers in a shallow unconfined aquifer: Journal of Hydrology, v. 103, p. 1–9.

Reimann, C., and de Caritat, P., 1998, Chemical elements in the environment, Factsheets for the Geochemist and Environmental Scientist: Berlin, Springer-Verlag, 398 p.

Rowe, B.L., Toccalino, P.L., Moran, M.J., Zogorski, J.S., and Price, C.V., 2007, Occurrence and potential human-health relevance of volatile organic compounds in drinking water from domestic wells in the United States: Environmental Health Perspectives, v. 115, no. 11, p. 1539–1546.

Saucedo, G.J., Bedford, D.R., Raines, G.L., Miller, R.J., and Wentworth, C.M., 2000, GIS data for the geologic map of California (version 2.0): Sacramento, Calif., California Department of Conservation, Division of Mines and Geology.

Schlosser, P., Stute, M., Dörr, H., Sonntag, C., and Munnich, K.O., 1988, Tritium/^{3}He dating of shallow groundwater: Earth and Planetary Science Letters, v. 89, p. 353–362.

Scott, J.C., 1990, Computerized stratified random site selection approaches for design of a ground-water quality sampling network: U.S. Geological Survey Water-Resources Investigations Report 90-4101, 109 p.

Shelton, J.L., Pimentel, Isabel, Fram, M.S., and Belitz, Kenneth, 2008, Ground-water quality data in the Kern County subbasin study unit, 2006—Results from the California GAMA Program: U.S. Geological Survey Data Series 337, 75 p. (Also available at http://pubs.usgs.gov/ds/337/.)

Solomon, D.K., and Cook, P.G., 2000, ^{3}H and ^{3}He, in Cook, P.G., and Herczeg, A.L., eds., Environmental tracers in subsurface hydrology: Boston, Kluwer Academic Press, p. 397–424.

Squillace, P.J., Moran, M.J., and Price, C.V., 2004, VOCs in shallow groundwater in new residential/commercial areas of the United States: Environmental Science and Technology, v. 38, no. 20, p. 5327–5338.

State of California, 1999, Resources—State Water Resources Control Board, in Supplemental Report of the 1999 Budget Act, 1999–00 Fiscal Year: State of California Legislative Analyst's Office website, Item 3940-001-0001, accessed March 16, 2011, at http://www.lao.ca.gov/1999/99-00_supp_rpt_lang.html#3940.

State of California, 2001a, Assembly Bill No. 599, Chapter 522, accessed March 16, 2011, at http://www.swrcb.ca.gov/gama/docs/ab_599_bill_20011005_chaptered.pdf.

State of California, 2001b, Groundwater Monitoring Act of 2001 (California Water Code, pt 2.76, sec. 10780-10782.3), accessed March 16, 2011, at http://www.leginfo.ca.gov/cgi-bin/displaycode?section=wat&group=10001-11000&file=10780-10782.3.

Stollenwerk, K.G., 2003, Geochemical processes controlling transport of arsenic in groundwater, a review of adsorption, in Welch, A.H., and Stollenwerk, K.G., eds., Arsenic in groundwater—geochemistry and occurrence: Boston, Kluwer Academic Publishers, 488 p.

Sturchio, N.C., Böhlke, J.K., Beloso, A.D., Streger, S.H., Heraty, L.J., and Hatzinger, P.B., 2007, Oxygen and chlorine isotopic fractionation during perchlorate biodegradation—Laboratory results and implications for forensics and natural attenuation studies: Environmental Science and Technology, v. 41, no. 8, p. 2796–2802.

Takaoka, N., and Mizutani, Y., 1987, Tritiogenic ^{3}He in groundwater in Takaoka: Earth and Planetary Science Letters, v. 85, p. 74–78.

Toccalino, P.L., and Norman, J.E., 2006, Health-based screening levels to evaluate U.S. Geological Survey ground water quality data: Risk Analysis, v. 26, no. 5, p. 1339–1348.

Toccalino, P.L., Norman, J.E., Phillips, R.H., Kauffman, L.J., Stackelberg, P.E., Nowell, L.H., Krietzman, S.J., and Post, G.B., 2004, Application of health-based screening levels to ground-water quality data in a state-scale pilot effort: U.S. Geological Survey Scientific Investigations Report 2004-5174, 64 p. (Also available at http://pubs.usgs.gov/sir/2004/5174/.)

Tolstikhin, I.N., and Kamensky, I.L., 1969, Determination of groundwater ages by the T-^{3}He method: Geochemistry International, v. 6, p. 810–811.

Torgersen, T., 1980, Controls on pore-fluid concentrations of ^{4}He and ^{222}Rn and the calculation of ^{4}He/^{222}Rn ages: Journal of Geochemical Exploration, v. 13, p. 57–75.

Torgersen, T., and Clarke, W.B., 1985, Helium accumulation in groundwater—I. An evaluation of sources and continental flux of crustal ^{4}He in the Great Artesian basin, Australia: Geochimica et Cosmochimica Acta, v. 49, p. 1211–1218.

Torgersen, T., Clarke, W.B., and Jenkins, W.J., 1979, The tritium/helium3 method in hydrology: International Atomic Energy Agency SM-228, v. 49, p. 917–930.

U.S. Census Bureau, 1990, U.S. Census ftp site, accessed September 28, 2010, at ftp://ftp2.census.gov/census_1990.

U.S. Census Bureau, 2000, Table 6—Population for the 15 largest counties and incorporated places in California, in Census 2000 Redistricting Data (P.L. 94-171) Summary file, Table PL1, and 1990 census, accessed March 28, 2011, at http://www.calinst.org/data/ca2kcens.pdf.

U.S. Environmental Protection Agency, 1997 Reporting Requirements for Risk/Benefit Information; Final Rule: Federal Register, v. 62, no. 182, Friday, September 19, 1997, Rules and Regulations, p. 49369–49395, accessed March 28, 2011, at http://www.gpoaccess.gov/fr/retrieve.html.

U.S. Environmental Protection Agency, 2000, National primary drinking water regulations; radionuclides; final rule, accessed February 28, 2012, at https://federalregister.gov/a/00-30421.

U.S. Environmental Protection Agency, 2006, 2006 Edition of the Drinking Water Standards and Health Advisories: U.S. Environmental Protection Agency, Office of Water EPA/822/R-06-013, accessed March 16, 2011, at http://www.epa.gov/waterscience/criteria/drinking/dwstandards.pdf.

U.S. Environmental Protection Agency, 2008a, Drinking water contaminants: U.S. Environmental Protection Agency website, accessed June 1, 2008, at http://www.epa.gov/safewater/contaminants/index.html.

U.S. Environmental Protection Agency, 2008b, Archived tables, 2006, *in* Drinking water health advisories and science support—Drinking water standards and health advisory tables: U.S. Environmental Protection Agency website, accessed June 1, 2008, at http://www.epa.gov/waterscience/criteria/drinking/.

U.S. Geological Survey, 2011, What is the Priority Basin Project?: U.S. Geological Survey website, accessed March 16, 2011, at http://ca.water.usgs.gov/gama.

Vine, J.D., and Tourtelot, E.B., 1970, Geochemistry of black shale deposits—a summary report. Economic Geology and the Bulletin of the Society of Economic Geologists, v. 65, no. 3, p. 253–272.

Vogel, J.C., and Ehhalt, D., 1963, The use of the carbon isotopes in groundwater studies, *in* Radioisotopes in Hydrology: Vienna, International Atomic Energy Agency, p. 383–395.

Water Education Foundation, 2006, Where does my water come from?: State Water Resources Control Board website, accessed March 16, 2011, at http://www.water-ed.org/watersources/default.asp.

Welch, A.H., Oremland, R.S., Davis, J.A., and Watkins, S.A., 2006, Arsenic in groundwater—A review of current knowledge and relation to the CALFED solution area with recommendations for needed research: San Francisco Estuary and Watershed Science, v. 4, no. 2, Article 4, 32 p.

Welch, A.H., Westjohn, D.B., Helsel, D.R., and Wanty, R.B., 2000, Arsenic in ground water of the United States—Occurrence and geochemistry: Ground Water, v. 38, no. 4, p. 589–604.

Wright, M.T., and Belitz, Kenneth, 2010, Factors controlling the regional distribution of vanadium in groundwater: Ground Water, v. 48, p. 515–525, doi:10.1111/j.1745-6584.2009.00666.x, accessed March 16, 2011, at http://onlinelibrary.wiley.com/doi/10.1111/j.1745-6584.2009.00666.x/abstract.

Wright, M.T., Belitz, K., and Johnson, T., 2004, Assessing the susceptibility to contamination of two aquifer systems used for public water supply in the Modesto and Fresno metropolitan areas, California, 2001 and 2002: U.S. Geological Survey Scientific Investigations Report 2004–5149, 35 p. (Also available at http://pubs.usgs.gov/sir/2004/5149/.)

Zebarth, B.J., Szeto, S.Y., Hii, B., Liebscher, H., and Grove, G., 1998, Groundwater contamination by chlorinated hydrocarbon impurities present in soil fumigant formulations: Water Quality Research Journal of Canada, v. 33, no. 1, p. 31–50.

Zogorski, J.S., Carter, J.M., Ivahnenko, T., Lapham, W.W., Moran, M.J., Rowe, B.L., Squillace, P.J., and Toccalino, P.L., 2006, The quality of our Nation's waters—Volatile organic compounds in the Nation's ground water and drinking-water supply wells: U.S. Geological Survey Circular 1292, 101 p. (Also available at http://pubs.usgs.gov/circ/circ1292/.)

Appendix A: Selection of California Department of Public Health Grid Wells

California requires samples to be collected regularly from public-supply wells under Title 22 (California Department of Health Services, 2007). Historical data derived from these samples are available from the California Department of Public Health (CDPH) database. Assembly Bill (AB) 599 directs the Groundwater Ambient Monitoring and Assessment (GAMA) Program to use available data and to collect new data as needed. The GAMA Priority Basin Project uses this existing monitoring data along with newly collected data to characterize the water quality of the primary aquifers. The CDPH database provided additional water-quality data for the spatially weighted and grid-based approaches to estimating aquifer-scale proportions for a wide range of constituents. CDPH data were not used to provide data for grid wells for VOCs, pesticides, or perchlorate because reporting limits for these constituents in the CDPH database generally were not sufficiently low enough to differentiate between "low" and "moderate" concentrations.

Three approaches were used to select CDPH inorganic constituent data for each grid cell where the USGS did not sample for inorganic constituents. The first approach was to identify CDPH data collected during the current period for the USGS-grid well (a well which was not sampled for inorganic constituents by the USGS). Analytical results were reviewed to determine if they met quality-control criteria to minimize the selection of poor-quality data from the CDPH database. Cation-anion balance was used as the quality-control assessment metric for selecting a CDPH-grid well. Because water is electrically neutral, the total positive charge on dissolved cation species in a water sample must equal the total negative charge on dissolved anion species; the cation/anion imbalance commonly is used as a quality-control criterion for water sample analysis (Hem, 1985). Cation-anion imbalance was calculated as the difference between the total cations and total anions divided by the sum, expressed as a percentage:

$$percent\ difference = \left(\frac{\left| \sum cations - \sum anions \right|}{\sum cations + \sum anions} \right) * 100$$

where

$\sum cations$ is the sum of calcium, magnesium, sodium, and potassium in milliequivalents per liter (meq/L), and

$\sum anions$ is the sum of chloride, sulfate, fluoride, nitrate, and bicarbonate in meq/L.

An imbalance, or percentage difference, greater than or equal to 10 percent indicates uncertainty in the quality of the data. The most recent CDPH data from the current period (January 1, 2003 to December 31, 2005) for the USGS-grid wells with missing data were evaluated to determine whether the cation/anion imbalance for CDPH data was less than 10 percent. If so, the CDPH inorganic data for the well was selected for use as the grid well data for inorganic constituents. It was assumed that if analyses met high-quality-control criteria for major-ion data, then the data quality for the analyses at these wells also would be acceptable for trace elements, nutrients, and radiochemical constituents. This approach resulted in the selection of inorganic data from CDPH at 30 USGS-grid wells in the SESJ study unit and 15 wells in the KERN study unit. For identification purposes, data from the CDPH for these grid wells were assigned GAMA identification numbers equivalent to the GAMA USGS-grid well number but with DG inserted between the study area prefix and sequence number (for example, CDPH-grid well KWH-DG-11 is the same well as USGS-grid well KWH-11, tables A1 and A2).

If the first approach did not yield CDPH inorganic data for a grid cell, the second approach was to search the CDPH database to identify the highest ranked well within that cell with a cation/anion imbalance of less than 10 percent. This approach resulted in selecting CDPH inorganic data for wells not sampled by USGS in 14 grid cells in the SESJ study unit and 21 grid cells in the KERN study unit. These 36 CDPH-grid wells were located within the same cell as the USGS-grid well but not necessarily right next to the USGS-grid well or in a grid cell not sampled by USGS-GAMA. To identify these new CDPH-grid wells, a well ID was created that added DPH after the study area prefix. If a USGS-grid well was sampled in that cell, the DPH was inserted between the study unit prefix and the sequence number (for example, CDPH-grid well KERN-DPH-05 is in the same cell as USGS-grid well KERN-05). If the CDPH-grid well was in a cell not sampled by USGS-GAMA then the DPH was followed by the next incremental number not used by the USGS-grid wells in that study area (for example KING-DPH-40).

If the cation-anion imbalance for data from the well in the CDPH database in a grid cell was not less than 10 percent, the third approach was to select the highest ranked well in the CDPH database with any of the needed inorganic data. This approach resulted in the selection of 15 USGS-grid wells in the SESJ study unit and 7 USGS-grid wells in the KERN study unit from which some CDPH inorganic data (usually nutrient data) were available. Because the wells were USGS-grid wells, a well ID was created that added DG to the GAMA ID (for example, well KWH-DG-13). In addition, this approach resulted in the selection of 9 CDPH-grid wells in the SESJ study unit and 14 CDPH-grid wells in the KERN study unit. The well ID for these CDPH-grid wells was assigned in the same manner as the other CDPH-grid wells.

The result of these approaches was one grid well per cell with data from the USGS database, the CDPH database, or both databases. Inorganic data for 128 CDPH-grid wells (68 in the SESJ study unit and 60 in the KERN study unit) in the CDPH database were used (fig. A1 and A2). Data were not available for all inorganic constituents from all 128 CDPH-grid wells. Table 4 in the report shows the number of USGS- and CDPH-grid wells with data for each inorganic constituent. In combination with USGS-grid-well inorganic data (42 wells), inorganic data was available for 148 grid cells. Most of the cells without a grid well were located in the area of the Tulare Lake bed or in the western part of the KERN study unit.

A larger error is associated with the 90 percent confidence intervals for estimates of aquifer-scale proportion for constituents made on the basis of a smaller number of wells (based on the Jeffreys interval for the binomial distribution, Brown and others, 2001). Analysis of the combined datasets to evaluate the occurrence of relatively high or moderate concentrations for inorganic constituents was not affected by differences in reporting levels between GAMA-collected and CDPH data because concentrations greater than one-half of water-quality benchmarks generally were substantially higher than the highest reporting levels. Comparisons between USGS-collected and CDPH data are described in appendix B.

Wells sampled by the Domestic Well Project are included in some of the analyses. Aquifer proportions were not calculated for these wells because they are not spatially distributed. Location and well identification are shown in figure A3.

Table A1.   Nomenclature and construction information for USGS-grid and USGS-understanding wells, and CDPH-grid wells, Southeast San Joaquin Valley study unit, California GAMA Priority Basin Project.

[**Abbreviations**: USGS, U.S. Geological Survey; GAMA, Groundwater Ambient Monitoring and Assessment Program; CDPH, California Department of Public Health; ft, foot; LSD, land-surface datum; KING, Kings study area well; KWH, Kaweah study area well; TLR, Tulare Lake study area well; TULE, Tule study area well; FP, flow-path well; HWY99T, transect well; DG, designates CDPH data from same well as USGS-grid well; DPH, designates well selected from subset of CDPH wells; na, not available]

USGS GAMA well identification number	CDPH GAMA well identification number	Cell number	Well construction information				
			Well type	Well depth, (ft below LSD)	Top of perforation (ft below LSD)	Bottom of perforation (ft below LSD)	Length of perforated interval (ft)
Grid wells							
KING-01	KING-DG-01	1	Production	610	580	600	20
KING-02	KING-DG-02	13	Production	na	na	na	na
KING-03	KING-DG-03	15	Production	380	145	358	213
KING-04		3	Production	500	240	500	260
KING-05	KING-DG-05	12	Production	555	505	545	40
KING-06		16	Production	705	345	695	350
KING-07	KING-DG-07	27	Production	410	194	410	216
KING-08	KING-DG-08	30	Production	540	287	540	253
KING-09	KING-DG-09	25	Production	582	202	380	178
KING-10		29	Production	228	228	228	0
KING-11	KING-DG-11	38	Production	540	280	520	240
KING-12	KING-DG-12	36	Production	246	126	246	120
KING-13	KING-DG-13	37	Production	420	260	400	140
KING-14	KING-DG-14	6	Production	800	640	780	140
KING-15	KING-DG-15	19	Production	440	270	440	170
KING-16	KING-DG-16	20	Production	na	na	na	na
KING-17		18	Production	650	320	640	320
KING-18	KING-DG-18	31	Production	370	160	360	200
KING-19	KING-DG-19	39	Production	260	218	260	42
KING-20		32	Production	124	na	na	na
KING-21	KING-DG-21	14	Production	210	150	210	60
KING-22	KING-DG-22	22	Production	510	280	500	220
KING-23	KING-DG-23	21	Production	540	150	510	360
KING-24	KING-DG-24	23	Production	236	140	236	96
KING-25	KING-DG-25	28	Production	384	168	384	216
KING-26		4	Production	480	240	480	240
KING-27	KING-DG-27	7	Production	409	na	na	na
KING-28	KING-DG-28	34	Production	76	40	74	34
KING-29	KING-DG-29	33	Production	390	120	390	270
KING-30		35	Production	490	330	470	140
KING-31		2	Production	520	280	520	240
KING-32	KING-DG-32	17	Production	445	370	445	75
KING-33	KING-DG-33	40	Production	675	175	655	480
KING-34	KING-DG-34	11	Production	na	na	na	na
KING-35	KING-DG-35	9	Production	na	na	na	na
KING-36	KING-DG-36	8	Production	452	na	na	na
KING-37		26	Production	na	na	na	na
KING-38		24	Production	700	na	na	na
KING-39		5	Production	na	na	na	na
KWH-01	KWH-DG-01	13	Production	640	245	620	375
KWH-02	KWH-DG-02	14	Production	225	116	225	109
KWH-03	KWH-DG-03	8	Production	310	na	na	na
KWH-04	KWH-DG-04	16	Production	205	80	200	120

Table A1. Nomenclature and construction information for USGS-grid and USGS-understanding wells, and CDPH-grid wells, Southeast San Joaquin Valley study unit, California GAMA Priority Basin Project.—Continued

[Abbreviations: USGS, U.S. Geological Survey; GAMA, Groundwater Ambient Monitoring and Assessment Program; CDPH, California Department of Public Health; ft, foot; LSD, land-surface datum; KING, Kings study area well; KWH, Kaweah study area well; TLR, Tulare Lake study area well; TULE, Tule study area well; FP, flow-path well; HWY99T, transect well; DG, designates CDPH data from same well as USGS-grid well; DPH, designates well selected from subset of CDPH wells; na, not available]

USGS GAMA well identification number	CDPH GAMA well identification number	Cell number	Well construction information				
			Well type	Well depth, (ft below LSD)	Top of perforation (ft below LSD)	Bottom of perforation (ft below LSD)	Length of perforated interval (ft)
Grid wells—Continued							
KWH-05	KWH-DG-05	15	Production	na	na	na	na
KWH-06	KWH-DG-06	19	Production	580	300	580	280
KWH-07	KWH-DG-07	17	Production	na	na	na	na
KWH-08	KWH-DG-08	18	Production	na	na	na	na
KWH-09	KWH-DG-09	12	Production	na	na	na	na
KWH-10	KWH-DG-10	20	Production	323	na	na	na
KWH-11	KWH-DG-11	10	Production	700	300	700	400
KWH-12		9	Production	404	205	381	176
KWH-13	KWH-DG-13	2	Production	220	na	na	na
KWH-14		7	Production	400	175	390	215
KWH-15	KWH-DG-15	3	Production	na	na	na	na
KWH-16		11	Domestic	240	200	240	40
KWH-17	KWH-DG-17	1	Production	350	225	338	113
KWH-18		6	Production	165	95	155	60
TLR-01	TLR-DG-01	21	Production	na	na	na	na
TLR-02	TLR-DG-02	13	Production	1,330	1,000	1,330	330
TLR-03	TLR-DG-03	15	Production	1,420	1,067	1,395	328
TLR-04		14	Production	1,320	980	1,300	320
TLR-05		5	Production	561	311	561	250
TLR-06	TLR-DG-06	6	Production	na	na	na	na
TLR-07	TLR-DG-07	2	Production	570	210	545	335
TLR-08		22	Production	na	na	na	na
TLR-09		4	Production	1,200	na	na	na
TULE-01		9	Production	800	400	800	400
TULE-02	TULE-DG-02	15	Production	na	na	na	na
TULE-03	TULE-DG-03	6	Production	280	169	270	101
TULE-04	TULE-DG-04	20	Production	820	550	800	250
TULE-05	TULE-DG-05	19	Production	1,368	930	1,348	418
TULE-06		18	Production	1,641	702	1,641	939
TULE-07	TULE-DG-07	7	Production	600	na	na	na
TULE-08		8	Production	810	340	810	470
TULE-09	TULE-DG-09	16	Production	587	160	587	427
TULE-10	TULE-DG-10	17	Production	965	201	· 965	764
TULE-11		1	Production	1,150	600	1,150	550
TULE-12		3	Production	na	na	na	na
TULE-13	TULE-DG-13	14	Production	400	120	400	280
TULE-14	TULE-DG-14	13	Production	600	280	600	320
TULE-15		2	Production	1,330	970	1,265	295
TULE-16		11	Domestic	na	na	na	na
TULE-17	TULE-DG-17	10	Production	245	185	240	55
	KING-DPH-06	16	Production	na	na	na	na
	KING-DPH-08	30	Production	na	na	na	na

Table A1. Nomenclature and construction information for USGS-grid and USGS-understanding wells, and CDPH-grid wells, Southeast San Joaquin Valley study unit, California GAMA Priority Basin Project.—Continued

[**Abbreviations**: USGS, U.S. Geological Survey; GAMA, Groundwater Ambient Monitoring and Assessment Program; CDPH, California Department of Public Health; ft, foot; LSD, land-surface datum; KING, Kings study area well; KWH, Kaweah study area well; TLR, Tulare Lake study area well; TULE, Tule study area well; FP, flow-path well; HWY99T, transect well; DG, designates CDPH data from same well as USGS-grid well; DPH, designates well selected from subset of CDPH wells; na, not available]

USGS GAMA well identification number	CDPH GAMA well identification number	Cell number	Well construction information				
			Well type	Well depth, (ft below LSD)	Top of perforation (ft below LSD)	Bottom of perforation (ft below LSD)	Length of perforated interval (ft)
Grid wells—Continued							
	KING-DPH-10	29	Production	na	na	na	na
	KING-DPH-14	6	Production	na	na	na	na
	KING-DPH-27	7	Production	na	na	na	na
	KING-DPH-28	34	Production	400	40	400	360
	KING-DPH-32	17	Production	na	na	na	na
	KING-DPH-38	24	Production	na	na	na	na
	KING-DPH-39	5	Production	580	310	570	260
	KING-DPH-40	2	Production	na	na	na	na
	KWH-DPH-07	17	Production	na	na	na	na
	KWH-DPH-08	18	Production	na	na	na	na
	KWH-DPH-09	12	Production	350	254	350	96
	KWH-DPH-10	20	Production	na	na	na	na
	KWH-DPH-17	1	Production	1,150	520	1,130	610
	KWH-DPH-19	4	Production	690	230	690	460
	KWH-DPH-20	5	Production	na	na	na	na
	KWH-DPH-21	11	Production	na	na	na	na
	TLR-DPH-08	22	Production	na	na	na	na
	TLR-DPH-09	4	Production	na	na	na	na
	TULE-DPH-06	18	Production	600	200	590	390
	TULE-DPH-10	17	Production	200	30	115	85
	TULE-DPH-16	11	Production	800	700	800	100
	TULE-DPH-18	4	Production	1,210	560	1,210	650
Understanding wells							
KINGFP-01		23	Production	320	160	310	150
KINGFP-02		23	Production	620	410	610	200
KINGFP-03		23	Production	550	445	540	95
KINGFP-04		23	Production	420	150	240	90
KINGFP-05		23	Monitoring well	172	162	167	5
KINGFP-06		23	Monitoring well	265	255	260	5
KINGFP-07		23	Monitoring well	70	60	65	5
KINGFP-08		32	Monitoring well	81	71	76	5
KINGFP-09		32	Monitoring well	168	158	163	5
KINGFP-10		32	Monitoring well	268	258	263	5
KINGFP-11		23	Monitoring well	158	148	153	5
KINGFP-12		23	Monitoring well	80	70	75	5
KINGFP-13		18	Monitoring well	148	108	138	30
KINGFP-14		9	Monitoring well	140	110	130	20
KINGFP-15		9	Monitoring well	208	178	198	20
HWY99T-01		20	Production	681	213	670	457

Table A2. Nomenclature and construction information for USGS-grid and USGS-understanding wells, and CDPH-grid wells, Kern County Subbasin study unit, California GAMA Priority Basin Project.

[**Abbreviations**: USGS, U.S. Geological Survey; GAMA, Groundwater Ambient Monitoring and Assessment Program; CDPH, California Department of Public Health; No., number; ft, foot; LSD, land-surface datum; KERN, Kern study unit well; FP, flow-path well; DG, designates well with CDPH data from same well as USGS-grid well; DPH, designates well selected from subset of CDPH wells; na, not available]

USGS GAMA well identification number	CDPH GAMA well identification number	Cell No.	Well construction information				
			Well type	Well depth, (ft below LSD)	Top of perforation (ft below LSD)	Bottom of perforation (ft below LSD)	Length of perforated interval (ft)
Grid wells							
KERN-01		89	Production	560	220	560	340
KERN-02		117	Production	702	402	702	300
KERN-03	KERN-DG-03	104	Production	755	400	na	355
KERN-04	KERN-DG-04	101	Production	637	na	na	na
KERN-05		27	Domestic	522	396	522	126
KERN-06	KERN-DG-06	36	Production	700	400	700	300
KERN-07	KERN-DG-07	78	Production	1,000	400	1,000	600
KERN-08	KERN-DG-08	50	Production	430	290	410	120
KERN-09	KERN-DG-09	48	Production	400	150	400	250
KERN-10	KERN-DG-10	87	Production	1,007	119	1,007	888
KERN-11	KERN-DG-11	81	Production	785	440	785	345
KERN-12	KERN-DG-12	40	Production	500	200	500	300
KERN-13	KERN-DG-13	52	Production	790	380	780	400
KERN-14	KERN-DG-14	82	Production	1,023	630	1,008	378
KERN-15	KERN-DG-15	103	Production	724	340	724	384
KERN-16		42	Production	460	160	460	300
KERN-17	KERN-DG-17	96	Production	627	na	na	na
KERN-18	KERN-DG-18	85	Production	740	na	na	na
KERN-19	KERN-DG-19	98	Production	700	260	680	420
KERN-20		70	Production	1,000	350	1,000	650
KERN-21		41	Production	455	125	455	330
KERN-22	KERN-DG-22	99	Production	970	502	970	468
KERN-23	KERN-DG-23	25	Production	na	na	na	na
KERN-24		39	Domestic	500	400	500	100
KERN-25		118	Domestic	900	639	900	261
KERN-26	KERN-DG-26	94	Production	400	320	400	80
KERN-27		51	Production	600	na	na	na
KERN-28	KERN-DG-28	83	Production	763	482	763	281
KERN-29	KERN-DG-29	79	Production	500	320	500	180
KERN-30	KERN-DG-30	77	Production	1,008	294	1,008	714
KERN-31	KERN-DG-31	102	Production	590	300	590	290
KERN-32	KERN-DG-32	90	Production	730	590	730	140
KERN-33		120	Production	793	395	793	398
KERN-34		62	Production	765	106	762	656
KERN-35		84	Production	680	240	na	na
KERN-36		97	Production	720	400	720	320
KERN-37		100	Production	500	na	na	na
KERN-38		116	Production	na	na	na	na
KERN-39		59	Domestic	694	494	694	200
KERN-40		110	Production	1,496	1,048	1,400	352
KERN-41		56	Production	1,390	955	1,370	415
KERN-42		95	Production	718	na	na	na
KERN-43	KERN-DG-43	108	Production	880	730	870	140
KERN-44		58	Production	709	300	709	409

Table A2. Nomenclature and construction information for USGS-grid and USGS-understanding wells, and CDPH-grid wells, Kern County Subbasin study unit, California GAMA Priority Basin Project.—Continued

[**Abbreviations**: USGS, U.S. Geological Survey; GAMA, Groundwater Ambient Monitoring and Assessment Program; CDPH, California Department of Public Health; No., number; ft, foot; LSD, land-surface datum; KERN, Kern study unit well; FP, flow-path well; DG, designates well with CDPH data from same well as USGS-grid well; DPH, designates well selected from subset of CDPH wells; na, not available]

USGS GAMA well identification number	CDPH GAMA well identification number	Cell No.	Well construction information				
			Well type	Well depth, (ft below LSD)	Top of perforation (ft below LSD)	Bottom of perforation (ft below LSD)	Length of perforated interval (ft)
Grid wells—Continued							
KERN-45		60	Production	600	na	na	na
KERN-46		107	Production	1,028	483	1,028	545
KERN-47	KERN-DG-47	73	Production	810	na	na	na
	KERN-DPH-01	89	Production	na	na	na	na
	KERN-DPH-02	117	Production	698	438	na	na
	KERN-DPH-05	27	Production	640	565	640	75
	KERN-DPH-16	42	Production	na	na	na	na
	KERN-DPH-20	70	Production	na	na	na	na
	KERN-DPH-21	41	Production	na	na	na	na
	KERN-DPH-24	39	Production	503	403	na	na
	KERN-DPH-25	118	Production	na	na	na	na
	KERN-DPH-27	51	Production	na	na	na	na
	KERN-DPH-28	83	Production	na	na	na	na
	KERN-DPH-30	77	Production	na	na	na	na
	KERN-DPH-31	102	Production	601	192	na	na
	KERN-DPH-33	120	Production	na	na	na	na
	KERN-DPH-46	107	Production	na	na	na	na
	KERN-DPH-48	38	Production	na	na	na	na
	KERN-DPH-49	49	Production	na	na	na	na
	KERN-DPH-50	53	Production	925	427	na	na
	KERN-DPH-51	57	Production	na	na	na	na
	KERN-DPH-52	61	Production	na	na	na	na
	KERN-DPH-53	63	Production	790	160	na	na
	KERN-DPH-54	69	Production	400	300	400	100
	KERN-DPH-55	72	Production	na	na	na	na
	KERN-DPH-56	74	Production	801	500	801	301
	KERN-DPH-57	75	Production	na	na	na	na
	KERN-DPH-58	76	Production	802	300	na	na
	KERN-DPH-59	80	Production	800	250	na	na
	KERN-DPH-60	91	Production	na	na	na	na
	KERN-DPH-61	93	Production	na	na	na	na
	KERN-DPH-62	105	Production	na	na	na	na
	KERN-DPH-63	106	Production	na	na	na	na
	KERN-DPH-64	111	Production	na	na	na	na
	KERN-DPH-65	112	Production	1,447	767	na	na
	KERN-DPH-66	114	Production	na	na	na	na
	KERN-DPH-67	119	Production	na	na	na	na
	KERN-DPH-68	121	Production	na	na	na	na
Understanding wells							
KERNFP-01		101	Production	750	400	740	340
KERNFP-02		97	Production	720	420	720	300
KERNFP-03		97	Production	477	220	457	237

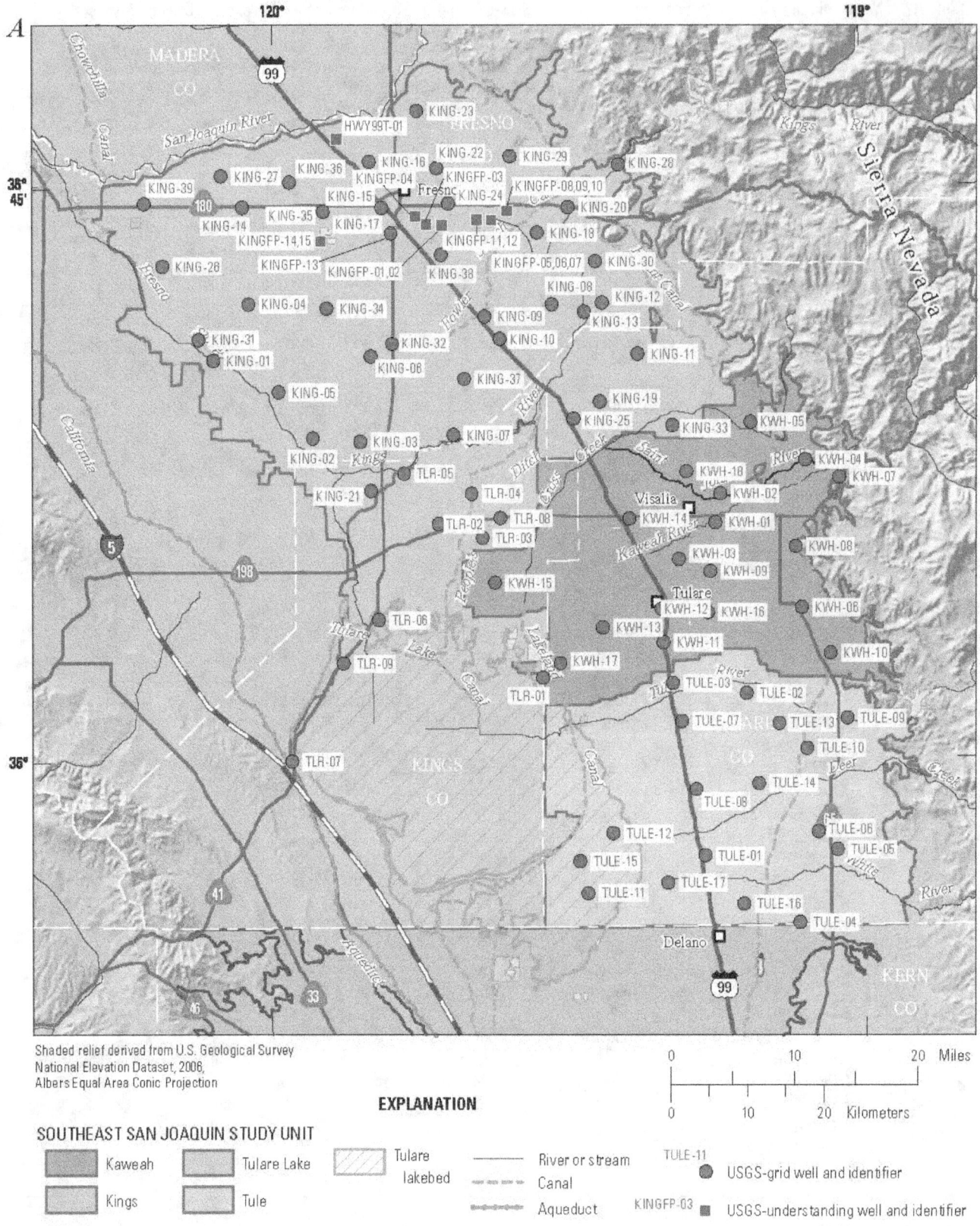

Figure A1. Identifiers and locations of (*A*) USGS-grid and USGS-understanding wells sampled during October 2005 through February 2006 and (*B*) CDPH-grid wells using data for inorganic constituents from the California Department of Public Health (CDPH), Southeast San Joaquin Valley study unit, southern San Joaquin Valley, California GAMA Priority Basin Project.

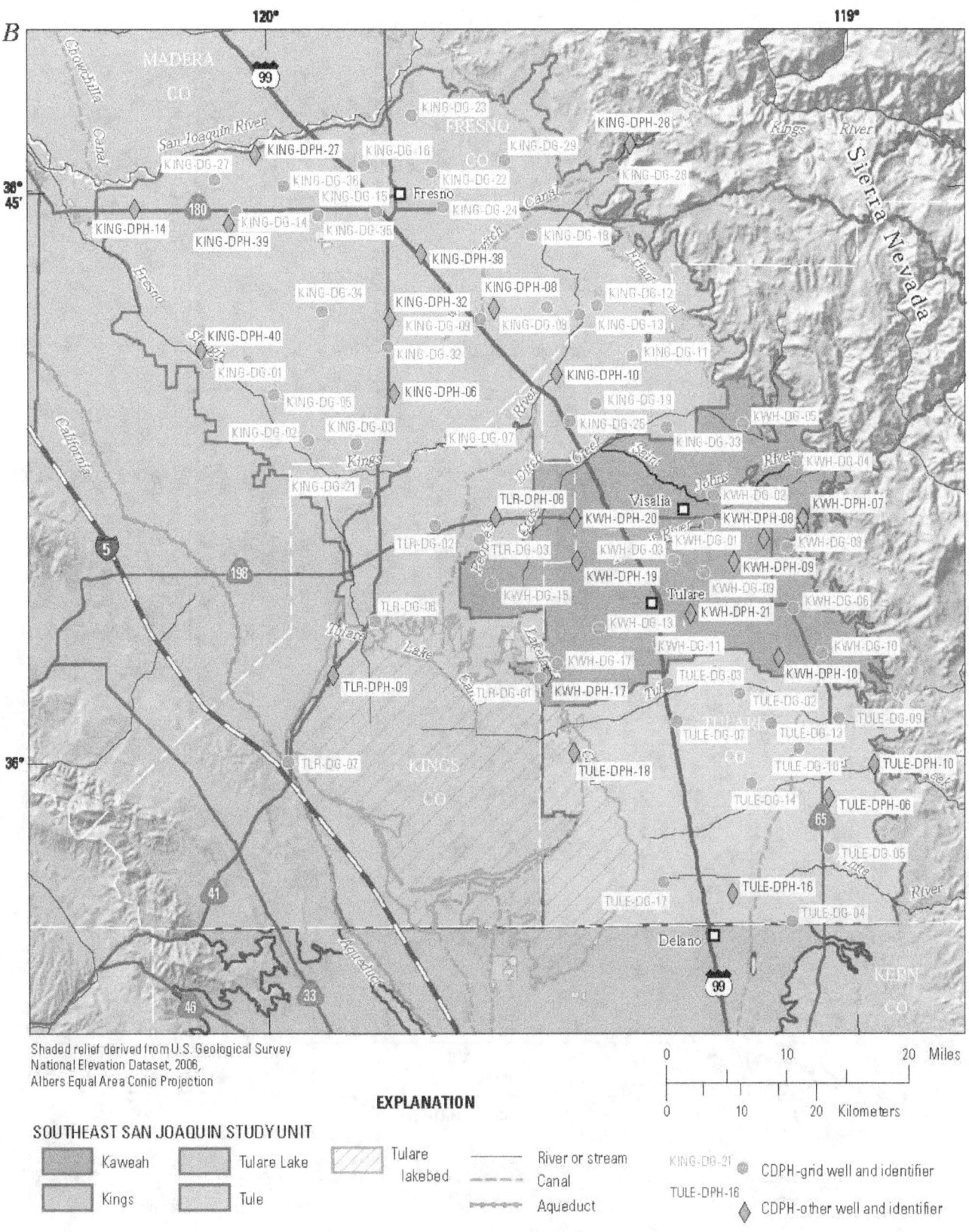

B

Shaded relief derived from U.S. Geological Survey
National Elevation Dataset, 2006,
Albers Equal Area Conic Projection

EXPLANATION

SOUTHEAST SAN JOAQUIN STUDY UNIT

▨ Kaweah	▨ Tulare Lake	▨ Tulare lakebed	—— River or stream
▨ Kings	▨ Tule		—·—·— Canal
			▸▸▸▸▸ Aqueduct

KING-DG-21 ● CDPH-grid well and identifier

TULE-DPH-16 ◆ CDPH-other well and identifier

Figure A1—Continued

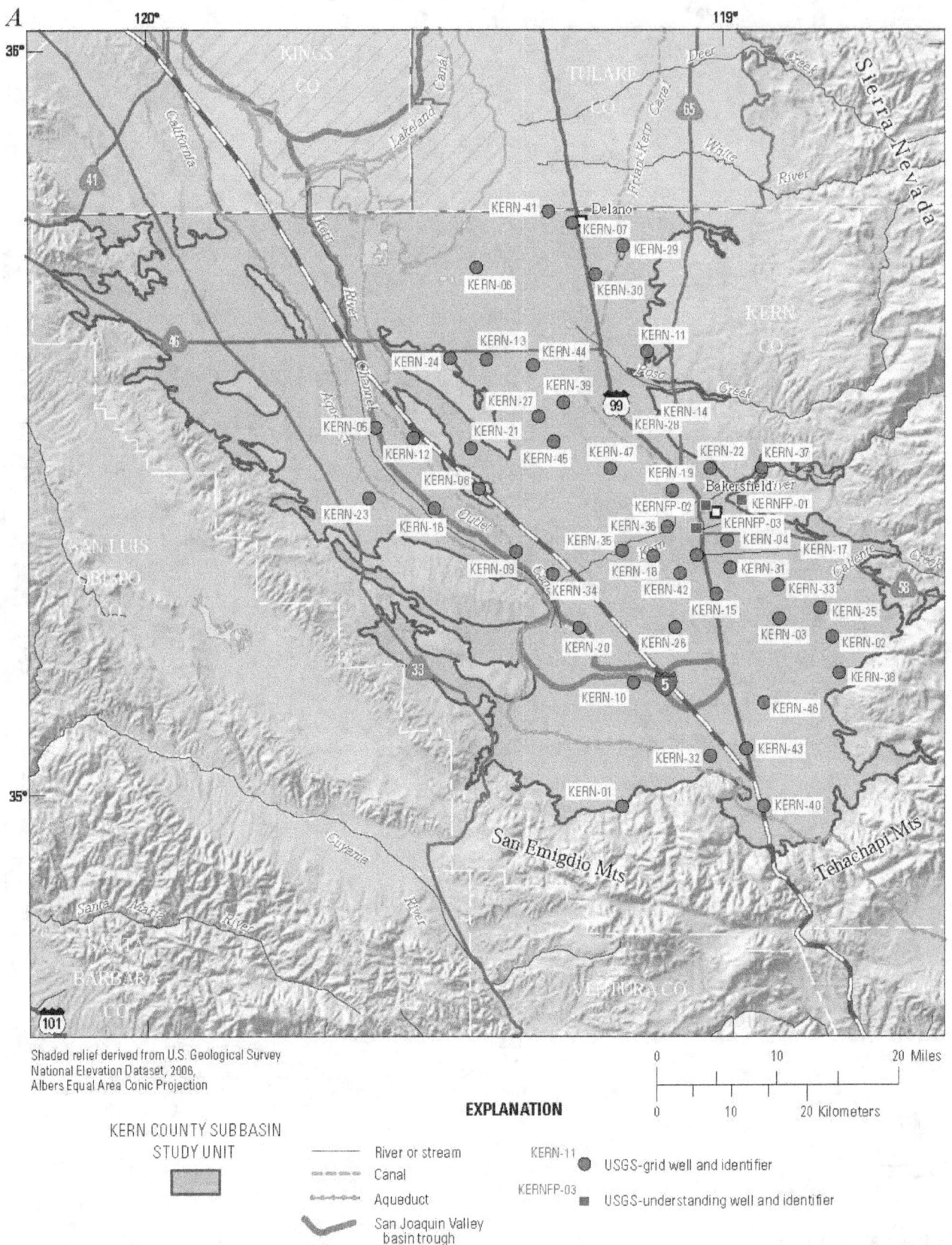

Shaded relief derived from U.S. Geological Survey
National Elevation Dataset, 2006,
Albers Equal Area Conic Projection

EXPLANATION

KERN COUNTY SUBBASIN
STUDY UNIT

— River or stream

- - - Canal

- - - Aqueduct

~ San Joaquin Valley
basin trough

KERN-11 ● USGS-grid well and identifier

KERNFP-03 ■ USGS-understanding well and identifier

Figure A2.   Identifiers and locations of (*A*) USGS-grid and USGS-understanding wells sampled during January through March 2006 and (*B*) CDPH-grid wells using data for inorganic constituents from the California Department of Public Health (CDPH), Kern County Subbasin study unit, southern San Joaquin Valley, California GAMA Priority Basin Project.

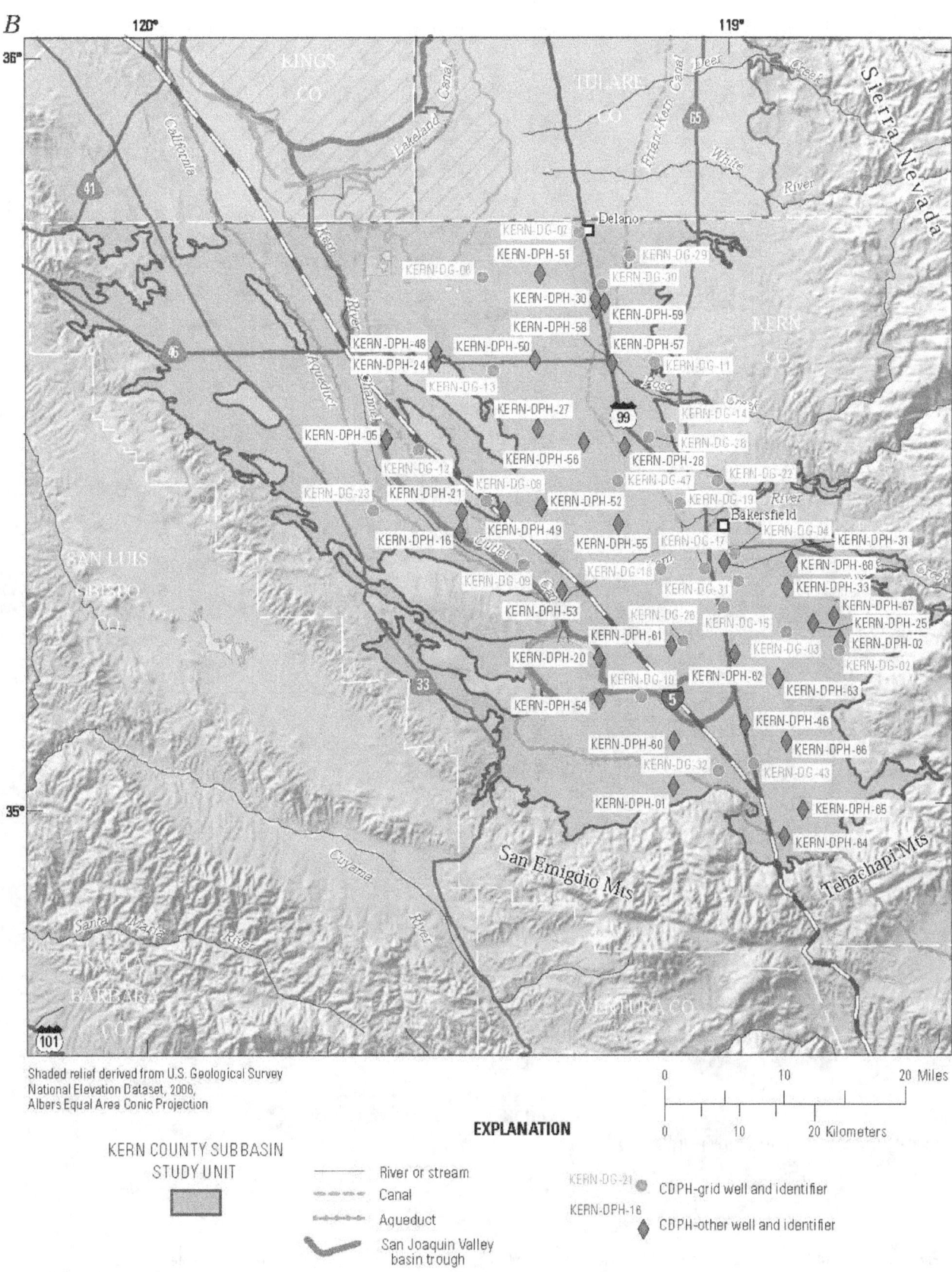

B

Shaded relief derived from U.S. Geological Survey
National Elevation Dataset, 2006,
Albers Equal Area Conic Projection

EXPLANATION

KERN COUNTY SUBBASIN
STUDY UNIT

——— River or stream

- - - Canal

Aqueduct

San Joaquin Valley
basin trough

KERN-DG-21 ● CDPH-grid well and identifier

KERN-DPH-16 ◆ CDPH-other well and identifier

0   10   20 Miles

0   10   20 Kilometers

Figure A2—Continued

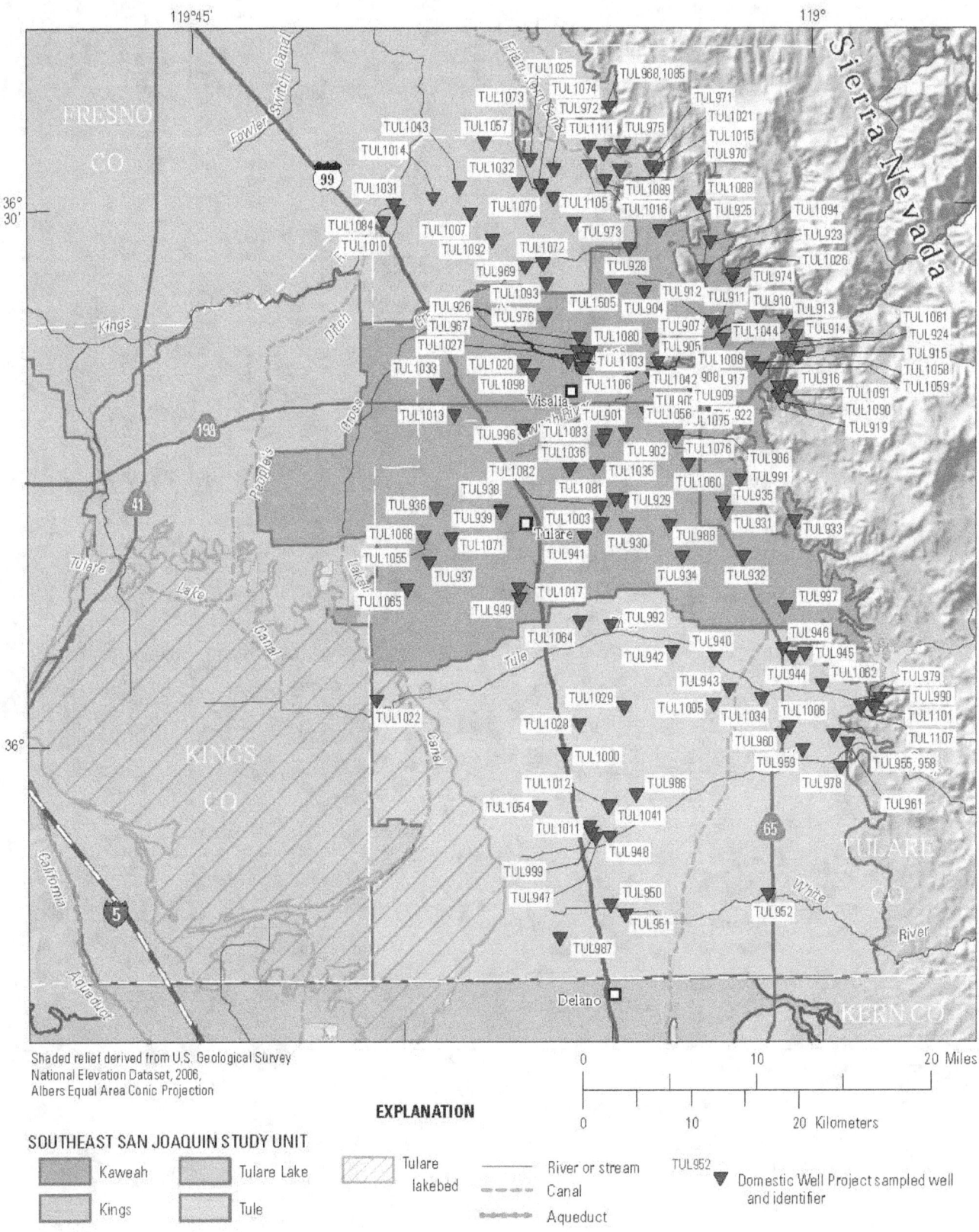

Figure A3. Identifiers and locations of domestic wells sampled by the California GAMA Domestic Well Project in the Southeast San Joaquin Valley study unit, southern San Joaquin Valley, California, 2006.

Appendix B: Comparison of CDPH and GAMA Priority Basin Data

CDPH and USGS-GAMA data were compared to assess the validity of combining data from these different sources. Because laboratory reporting levels for most organic constituents and trace elements were substantially lower for GAMA Priority Basin Project data than for CDPH data (table 2), it was not possible to directly compare concentrations of many constituents in individual wells in any meaningful way. However, concentrations of major ions and nitrate, which generally are prevalent and have concentrations substantially above reporting levels, could be compared for each well using data from both sources.

Comparisons were made for wells that were analyzed by USGS-GAMA Priority Basin Project for inorganic and radiochemical constituents and for which CDPH data were available within the most recent 3-year interval. Major ion, nitrate, and trace element data were available for 25 wells from the SESJ study unit and 14 wells from the KERN study unit in the USGS and the CDPH databases. Compared in the SESJ study unit were 22 to 25 pairs of data for nine constituents (calcium, magnesium, sodium, alkalinity, chloride, sulfate, total dissolved solids, nitrate, and arsenic). Three additional constituents were compared although there were fewer than 20 paired results for analysis (potassium, 17 pairs; fluoride, 15 pairs; and vanadium, 7 pairs). Compared in the KERN study unit were 10 to 14 pairs of data for 10 constituents (the same 9 as listed for the SESJ study unit plus fluoride). The dataset was large enough for meaningful statistical comparison for each constituent. Two additional constituents were compared although there were fewer than 10 paired results for analysis (potassium, 9 pairs; and vanadium, 6 pairs). Comparison tests performed on the combined datasets for the study units produced similar results to the individual study units.

A non-parametric signed-rank test indicated no significant differences between the paired USGS-GAMA and CDPH data for either the SESJ study unit (p-values ranging from 0.409 for alkalinity to 1.000 for magnesium) or the KERN study unit (p-values ranging from 0.507 for sodium to 0.982 for nitrate). Although differences between the paired datasets occurred for a few wells, most sample pairs plotted close to a 1-to-1 line (fig. B1). The relative percent difference (absolute difference of the two values divided by the average of the two values, RPD) was calculated for each data pair. The median RPD was less than 20 percent for all constituents except for vanadium in both study units. The median RPD for vanadium was 24 percent in the KERN study unit. These direct comparisons indicated that the GAMA and CDPH inorganic data were not significantly different.

Piper diagrams show the relative abundance of major cations and anions (on a charge equivalent basis) as a percentage of the total ion content of the water (fig. B2). Piper diagrams often are used to define groundwater type (Hem, 1985). Combined GAMA Priority Basin Project and CDPH major-ion data for grid wells were plotted on Piper diagrams (Piper, 1944) along with all CDPH major-ion data from the current period to determine whether the groundwater types in grid wells were similar to groundwater types observed historically in the study unit. All cation/anion data in the CDPH database with a cation/anion balance less than 10 percent were retrieved and plotted on these Piper diagrams for comparison with grid well data.

The range of water types for grid wells and other wells from the CDPH database for the current period were similar (fig. B2A and B2B). Most wells in the SESJ study unit were classified as *mixed cation-bicarbonate* type waters, indicating that no single cation accounted for more than 60 percent of the total cations, and bicarbonate accounted for more than 60 percent of the total anions. Many of the wells were classified as *mixed cation-mixed anion* type waters, indicating that no single cation accounted for more than 60 percent of the total cations, and no single anion accounted for more than 60 percent of the total anions. The most common cations were calcium and sodium, although some samples also contained a high percentage of magnesium. Bicarbonate and chloride were the dominant anions in these waters.

About one-third of the wells in the KERN study unit were classified as *mixed cation-bicarbonate*, where the dominant cations are calcium and sodium. The remaining wells were classified as *mixed cation-mixed anion*. In comparison with the SESJ study unit, sulfate was the dominant cation in more wells in the KERN study unit (fig. B2).

The determination that the range of relative abundance of major cations and anions in grid wells (83 in the SESJ study unit and 47 in the KERN study unit) is similar to the range of those in the selected CDPH-other wells (578 wells in the SESJ study unit and 263 in the KERN study unit) indicates that the grid wells represent the diversity of water types present within the southern San Joaquin Valley.

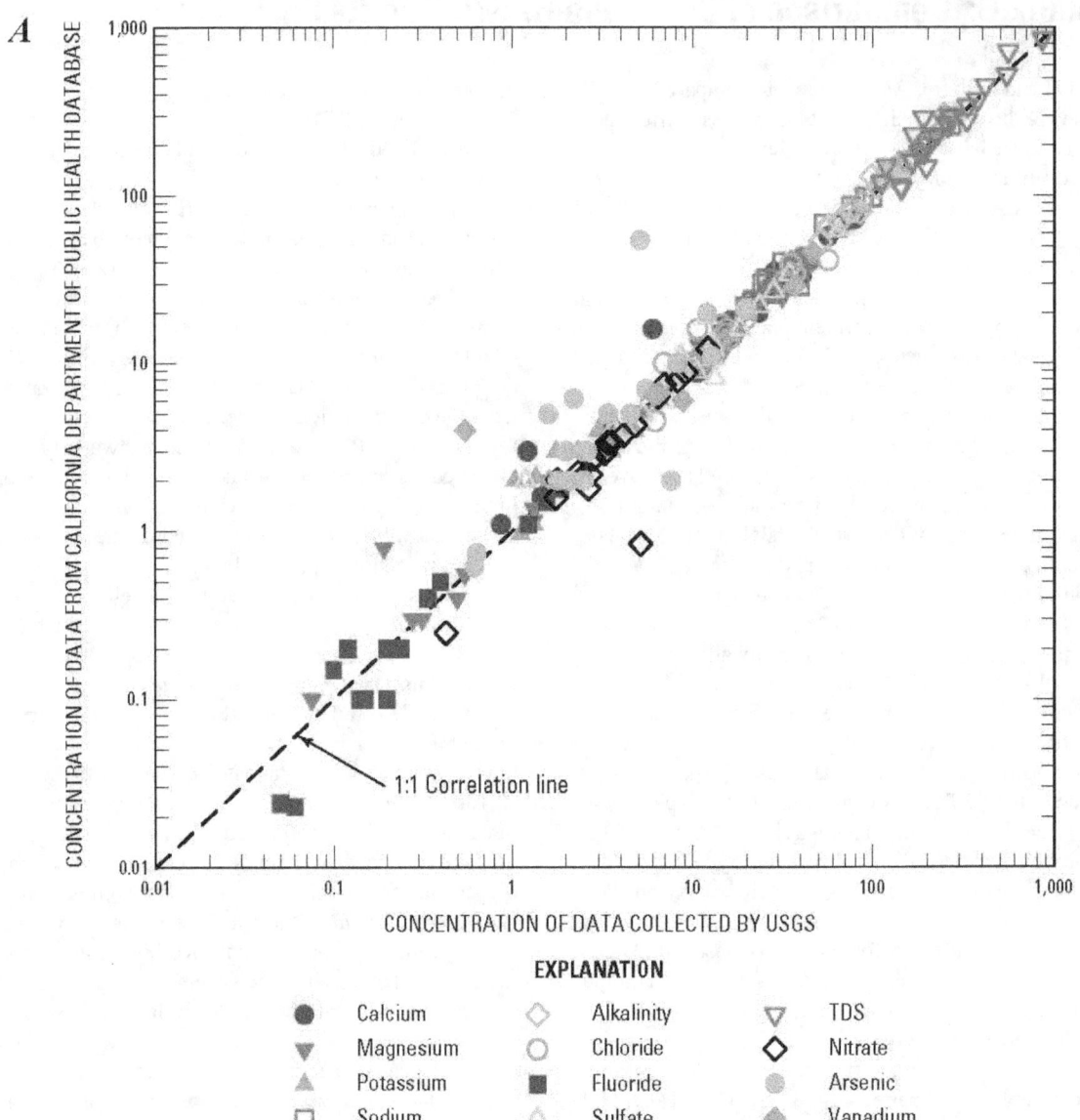

Figure B1. Paired inorganic concentrations from wells sampled in the southern San Joaquin Valley (October 2005 to March 2006) and the most recent available analysis in the California Department of Health Services (January 1, 2003, to December 31, 2005) in the (*A*) Southeast San Joaquin Valley study unit and the (*B*) Kern County Subbasin study unit, southern San Joaquin Valley, California GAMA Priority Basin Project.

B

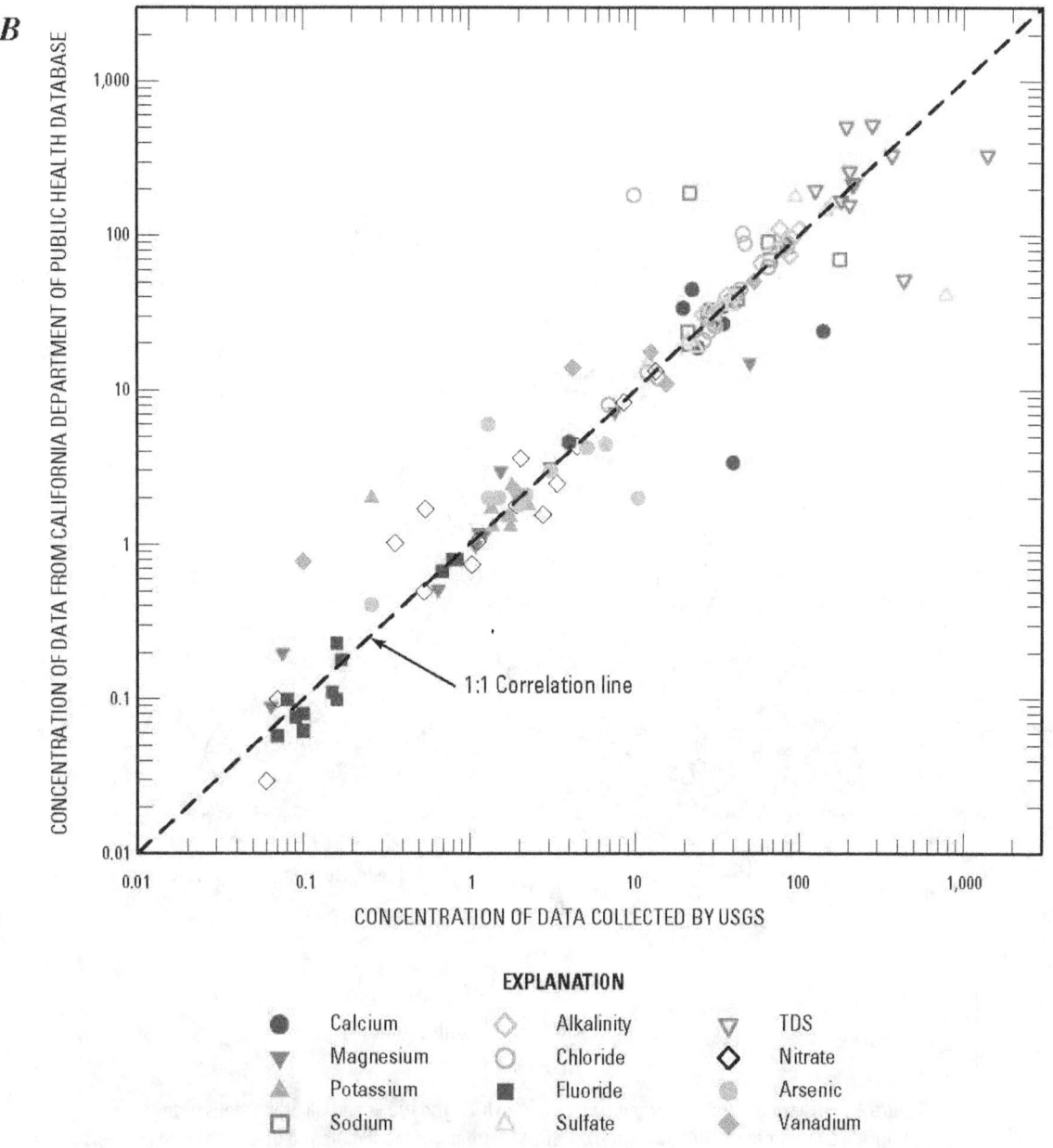

Figure B1—Continued

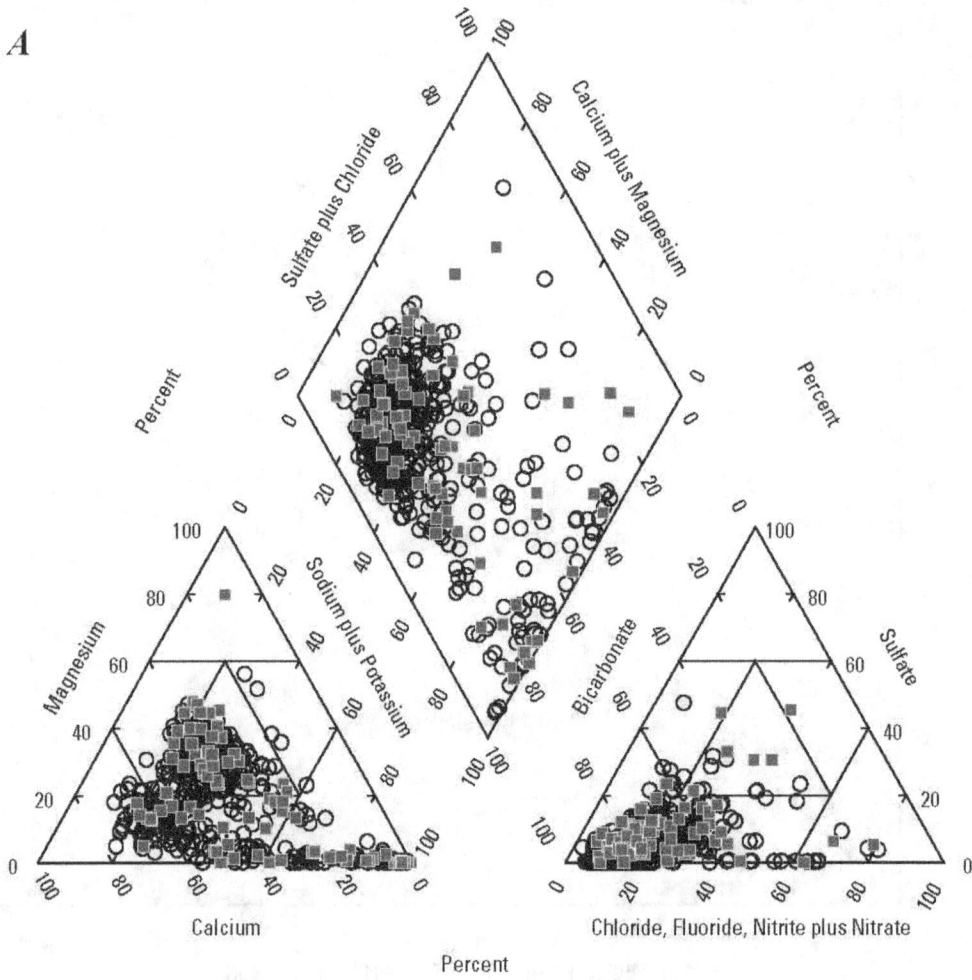

EXPLANATION

■ USGS-grid well

○ CDPH well

Figure B2.   Piper diagram showing USGS- and CDPH-grid wells and all other wells in the California Department of Public Health database with a charge imbalance of less than 10 percent in the (*A*) Southeast San Joaquin Valley study unit and in the (*B*) Kern County Subbasin study unit, southern San Joaquin Valley, California GAMA Priority Basin Project.

B

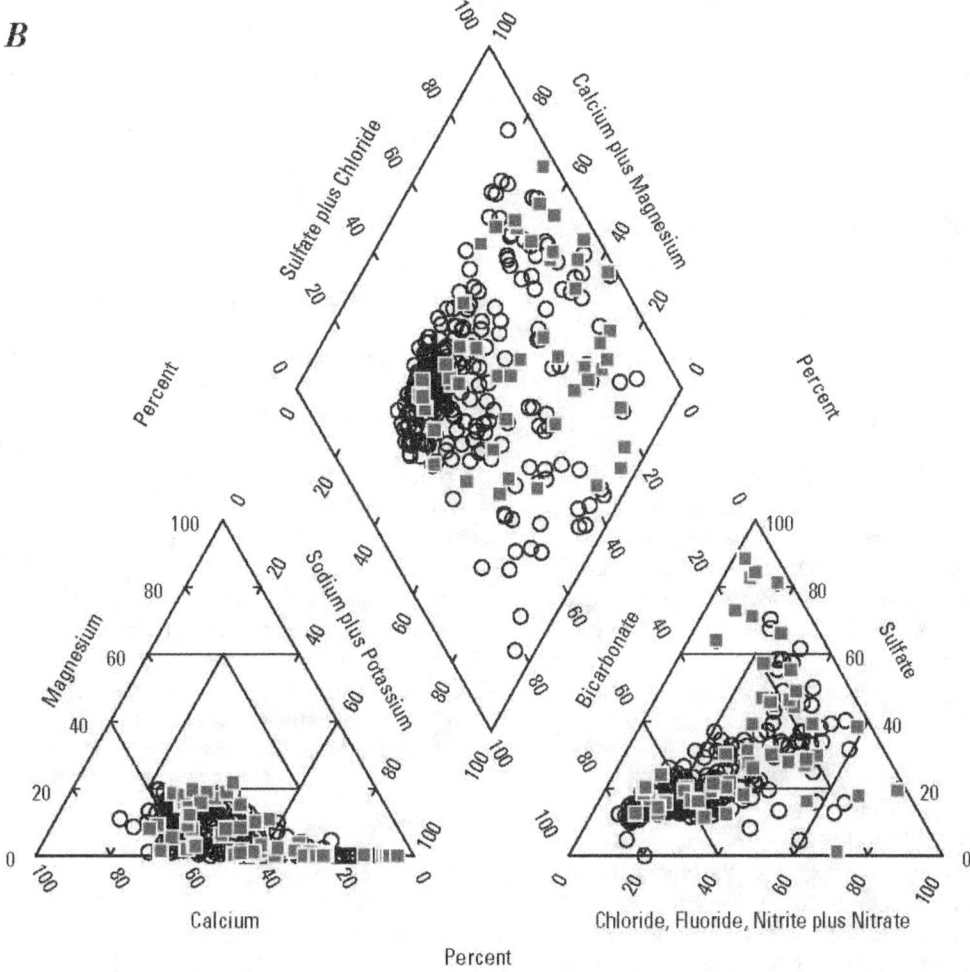

EXPLANATION

■ USGS-grid well

○ CDPH well

Figure B2—Continued

Appendix C: Calculation of Aquifer-Scale Proportions

Two statistical approaches—grid-based and spatially weighted—were selected to evaluate the aquifer-scale proportions of the primary aquifers in the southern San Joaquin Valley study units with high, moderate, or low relative-concentrations (concentration relative to its water-quality benchmark) of constituents. Raw detection frequencies also were calculated for individual constituents, but were not used for estimating aquifer-scale proportions because this method creates spatial bias toward regions with large numbers of wells.

Grid-Based Calculation

One well in each grid cell, a "grid well," was used to represent the primary aquifers. Most grid wells sampled for this study were USGS-grid wells. The relative-concentration for each constituent (concentration relative to its water-quality benchmark) was then evaluated for each grid well. The proportion of the primary aquifers with high relative-concentrations was calculated by dividing the number of cells with concentrations greater than the benchmark (relative-concentration greater than 1) by the total number of grid wells in each study unit (Belitz and others, 2010). Proportions containing moderate and low relative-concentrations were calculated similarly. Confidence intervals for grid-based aquifer-scale proportions were computed using the Jeffreys interval for the binomial distribution (Brown and others, 2001). The grid-based estimate is spatially unbiased. However, the grid-based approach may not detect constituents that are present at high relative-concentrations in small proportions of the primary aquifers.

The grid-based aquifer-scale proportions for constituent classes also are calculated on a one-value-per-cell basis. A cell with a high relative-concentration for any constituent in the class is defined as a high cell, and the high proportion is the number of high cells divided by the number of cells with data for any of the constituents in that class. The moderate proportion for the constituent class is calculated similarly, except that a cell already defined as high cannot also be defined as moderate. A cell with a moderate relative-concentration for any constituent in the class without a high value for any constituent in the class is defined as moderate. The grid-based aquifer-scale proportion for the low category was calculated similarly, such that a cell could only be low if the relative-concentration was neither moderate nor high for any constituent in the class. The proportions for the high, moderate, and low categories were expected to total 100 percent, except for small differences as a result of rounding.

Spatially Weighted Calculation

The spatially weighted calculation of aquifer-scale proportions uses the most recent value for a constituent from all wells in the CDPH database with data in the 3-year interval prior to USGS-GAMA sampling (January 1, 2003, to December 31, 2005) in the study units, from all USGS-grid well data, and from selected USGS-understanding well data. The spatially weighted approach computes the aquifer-scale proportion using the percentage of wells with high relative-concentrations from all of the wells in each cell, instead of using data from only one well (Belitz and others, 2010). For each constituent, the high aquifer-scale proportion was computed by calculating the proportion of wells with high relative-concentrations in each grid-cell and dividing by the number of cells (Belitz and others, 2010):

$$P_i = \frac{W_{high}}{W_{total}}$$

$$P_{SU} = \frac{\sum_{i=1}^{n} P_i}{n}$$

where
P_i is the proportion of wells in the ith cell with high relative-concentrations,

W_{total} is the number of wells in the ith cell with data for the constituent,

W_{high} is the number of wells in the ith cell with high relative-concentratons,

P_{SU} is the aquifer-scale proportion for the study unit, and

n is the number of cells with data for the constituent.

Similar procedures were used to calculate the proportions of moderate and low relative-concentrations. The resulting proportions are spatially unbiased (Isaaks and Srivastava, 1989).

Raw Detection Frequencies

The raw detection frequencies of wells with high relative-concentrations for constituents were calculated using the same data used for the spatially weighted approach. Raw detection frequency is the percentage (frequency) of wells in the study unit with high relative-concentrations. However, raw detection frequencies are not spatially unbiased because the wells in the CDPH database and USGS-understanding wells are not uniformly distributed. Consequently, high relative-concentrations for wells clustered in a particular area represent a small part of the primary aquifers and could be given a disproportionately high weight compared to that given by spatially unbiased approaches. Raw detection frequencies of high relative-concentrations are provided for reference in this report but were not used to assess aquifer-scale proportions.

Appendix D: Attribution of Potential Explanatory Factors

Well Construction Information

Well construction data were from driller's logs or from information provided by the well owner. Well identification verification procedures are described by Burton and Belitz (2008) and Shelton and others (2008). Well depths and depths to the top and bottom of the perforated interval for USGS-grid, USGS-understanding, and CDPH-grid wells are listed for the SESJ study unit in table A1 and for the KERN study unit in table A2. Wells were classified as production wells, monitoring wells, or domestic wells. Production wells pump groundwater from the aquifer to a distribution system. Monitoring wells are short-screened wells installed exclusively for monitoring purposes. Domestic wells pump groundwater from the aquifer for home use. Well construction data for the domestic wells collected by SWRCB as part of GAMA's Domestic Well Project were obtained from the well owners and are shown in table D1.

Land-Use Classification

Land use was classified using an enhanced version of the satellite-derived (30-m pixel resolution) nationwide USGS National Land Cover Dataset (Nakagaki and others, 2007). This dataset has been used in previous national and regional studies relating land use to water quality (Gilliom and others, 2006; Zogorski and others, 2006). The dataset characterizes land cover during the early 1990s. The imagery is classified into 25 land-cover classifications (Nakagaki and Wolock, 2005). These 25 land-cover classifications were assigned to 3 general land-use classifications—urban, agricultural, and natural. Land-use statistics for each study unit and for circles with a radius of 500 m around each well in the SESJ study unit (table D2) and in the KERN study unit (table D3) and the wells from the Domestic Well Project (table D4) were assigned using USGS National Land Cover Dataset (Johnson and Belitz, 2009).

Lateral Position

The lateral position of wells within the valley serves as a proxy for the horizontal position in the regional groundwater-flow system. Regionally, groundwater primarily flows from the eastern boundary of the valley-fill deposits along the Sierra Nevada mountain front toward the southwest to the western boundary of the flow system (fig. 5). The groundwater-flow system has vertical flow components as well as horizontal flow components that deviate from the regional northeast-to-southwest flow direction in response to withdrawals and recharge related to groundwater development

for irrigation since the early to mid-1900s (Burow and others, 2008a, 2008b; California Department of Water Resources, 2008). These vertical and nonparallel horizontal flow components are superimposed on the topographically driven regional flow system. The aquifer system also contains large quantities of groundwater that was recharged before the modern flow system developed; under predominantly natural conditions, groundwater primarily had moved from northeast to southwest.

The normalized lateral position (hereinafter, lateral position) was calculated as part of a regional groundwater-flow modeling study for a set of 30 × 30-m-wide cells in the San Joaquin Valley (Faunt and others, 2009). Lateral positions were assigned to wells residing in those cells using ArcGIS (version 9.2). The lateral position of each well was calculated as the ratio of (1) the distance from the well to the San Joaquin Valley trough and (2) the total distance from the San Joaquin Valley trough to the eastern boundary of the valley. The eastern boundary of the valley was represented by the eastern boundary of the valley fill deposits and was assigned a value of 1. The valley trough was assigned a value of 0. Because the Tulare Lake bed and other dry lake beds in the southern KERN study unit are topographic lows covering large areas, the approximate boundary of these lake beds, where present, were used to represent the location of the valley trough (fig. 8). The San Joaquin Valley trough and the eastern boundary were represented as approximate line segments, and lateral position was calculated along lines perpendicular to both bounding lines. High values of lateral position indicate locations in the upgradient or proximal portion of the flow system, and lower values of lateral position indicate locations in the downgradient or distal portion of the flow system. Plotting of data, with respect to lateral position, also allows for aggregation of areally distributed data into a single, diagrammatic cross section across the study unit. Values for lateral position for each USGS-grid and USGS-understanding well, CDPH-grid well, and Domestic Well Project well are given in tables D2, D3, and D4, respectively.

Septic Systems

Septic tank density was determined from housing characteristics data from the 1990 U.S. Census (U.S. Census Bureau, 1990, ftp://ftp2.census.gov/census_1990). The density of septic tanks in each housing census block was calculated from the number of tanks and block area. The density of septic tanks around each well was calculated from the area-weighted mean of the block densities for blocks within a 500-m buffer around the well location (Tyler Johnson, U.S. Geological Survey, written commun., 2009) (tables D2, D3, and D4).

Formerly Leaking Underground Fuel Tanks

Density for formerly leaking underground fuel tanks (LUFTs) was determined from data obtained from the Geographic Information Management System GeoTracker (California Environmental Protection Agency, 2001). The density is a measure of the number of tanks in a Thiessen polygon in square kilometers (km^2). The boundaries of the Thiessen polygons are created by bisecting the distance between all surrounding LUFTs. For instance, if a tank is surrounded by four tanks each 1,000 m away, then the Thiessen polygon will be drawn exactly one-half of the distance to each tank (500 m), resulting in a polygon that is relatively small and therefore of high density. The density is calculated by dividing the number of tanks at a single location (usually one) and dividing it by the total area of the polygon. If the nearest tanks are many miles away, then the polygon will be large, and therefore the density will be relatively low. This measure was added because two wells could be each 100 m away from a LUFT, but one could be surrounded by 10 nearby tanks and the other secluded without another tank for 100 mi. The Thiessen polygon method is a non-interpolated measure of density that has the added value of being able to handle extreme high and low densities equally well. LUFT density data for each USGS-grid and USGS-understanding well and CDPH-grid well are in tables D2 and D3.

Groundwater Age Classification

Groundwater dating techniques indicate the time after the groundwater was last in contact with the atmosphere (residence time). Techniques used to estimate groundwater residence times or 'age' include those based on tritium (^{3}H; Tolstikhin and Kamensky, 1969; Torgersen and others, 1979) and ^{3}H in combination with its decay product helium-3 (^{3}He) (Schlosser and others, 1988), carbon-14 (^{14}C) activities (Vogel and Ehhalt, 1963; Plummer and others, 1993; Kalin, 2000), and dissolved noble gases, particularly helium-4 accumulation (Davis and DeWiest, 1966; Andrews and Lee, 1979; Cey and others, 2008; Kulongoski and others, 2008).

Tritium (^{3}H) is a short-lived radioactive isotope of hydrogen with a half-life of 12.32 years (Lucas and Unterweger, 2000). Tritium is produced naturally in the atmosphere from the interaction of cosmogenic radiation with nitrogen (Craig and Lal, 1961), by above-ground nuclear explosions, and by the operation of nuclear reactors. Above-ground nuclear-bomb testing between 1951 and 1980 (peak production in 1963) introduced much larger quantities of ^{3}H than natural production into the atmosphere (Michel, 1989; Solomon and Cook, 2000). Tritium enters the hydrologic cycle as precipitation following oxidation to tritiated water.

Consequently, the presence of ^{3}H in groundwater may be used to identify water that has exchanged with the atmosphere in the past 50 years. By determining the ratio of ^{3}H to its decay product ^{3}He, the time that the water has resided in the aquifer can be calculated more precisely than using ^{3}H alone (Takaoka and Mizutani, 1987; Poreda and others, 1988). Tritium activity and tritium-helium age of the water samples are shown in tables D5 and D6.

^{14}C is a radioactive isotope of carbon, with a half-life of 5,730 years, that is formed naturally in the atmosphere by the interaction of cosmic-ray neutrons with nitrogen, and to a lesser degree, interaction with oxygen and carbon. ^{14}C is incorporated into carbon dioxide which is mixed throughout the atmosphere. Carbon dioxide is dissolved in precipitation and incorporated into the hydrologic cycle. ^{14}C activity in groundwater, expressed as percent modern carbon (pmc), indicates exposure to the atmospheric ^{14}C source. ^{14}C can be used to estimate groundwater ages ranging from 1,000 to less than 30,000 years before present (Clark and Fritz, 1997). Calculated ^{14}C ages (tables D5 and D6) in this study are referred to as "uncorrected" because they have not been adjusted to consider exchanges with sedimentary sources of carbon (Fontes and Garnier, 1979; Kalin, 2000).The ^{14}C age (residence time) is calculated based on the decrease in ^{14}C activity as a result of radioactive decay after groundwater recharge, relative to an assumed initial ^{14}C concentration (Clark and Fritz, 1997). A mean initial ^{14}C activity of 99 pmc was assumed for this study, with estimated errors on calculated groundwater ages as great as ±20 percent.

Helium (He) is a naturally occurring inert gas produced by the radioactive decay of lithium, thorium, and uranium in the Earth. Measured He concentrations in groundwater is the sum of air-equilibrated He, He from dissolved-air bubbles, terrigenic He, and tritiogenic ^{3}He. The helium (^{3}He and ^{4}He isotopes) concentrations in groundwater often exceed the expected solubility equilibrium values, which are a function of the temperature of the water, as a result of subsurface production of both isotopes, and their subsequent release into the groundwater (Morrison and Pine, 1955; Andrews and Lee, 1979; Torgersen, 1980; Andrews, 1985; Torgersen and Clark, 1985). The presence of terrigenic He in groundwater, from its production in aquifer material or deeper in the crust, is indicative of long groundwater residence times. The amount of terrigenic He is defined as the concentration of the total measured He, minus He from air equilibration and dissolved-air bubbles. Percentage of terrigenic He is defined as the concentration of terrigenic He (as defined previously) divided by the total measured He in the sample (corrected for air-bubble entrainment). Samples in which more than 5 percent of the total He is terrigenic He (percentage of terrigenic He) indicate groundwater has a residence time of more than 100 years.

Recharge temperatures for 138 samples were determined from dissolved neon, argon, krypton, and xenon data using methods described by Aeschbach-Hertig and others (1999). The only modeled recharge temperatures accepted were those for which the probability was greater than 1 percent that the sum of the squared deviations between the modeled and the measured concentrations (weighted with the experimental 1-sigma errors) was equal to or greater than the observed value (Aeschbach-Hertig and others, 2000).

The groundwater age was computed using ^{3}H /^{3}He as described by Poreda and others (1988). The ^{3}He/^{4}He of samples was determined by the linear regression of the percentage of terrigenic He and δ^3He ([δ^3He = R_{meas}/R_{atm} -1] $\times$ 100) of samples with less than 1 tritium unit (TU). R_{meas} is the ratio of ^{3}He/^{4}He in the measured sample; R_{atm} is the ratio of ^{3}He/^{4}He in the atmosphere. Calculations of the recharge temperature using noble gases and ^{3}He/^{4}He are useful because they can be used to constrain He-based groundwater ages further.

In this study, the ages of samples are classified as pre-modern, modern, and mixed (tables D5 and D6). Groundwater with ^{3}H activity less than 1 TU, percentage of terrigenic He greater than 5 percent, and ^{14}C less than 90 pmc was designated as pre-modern. Pre-modern groundwater is defined as having been recharged before 1952. Groundwater with ^{3}H greater than 1 TU, percentage of terrigenic He less than 5 percent, and ^{14}C greater than 90 pmc is designated as modern. Modern groundwater is defined as having been recharged after 1952. Samples with pre-modern and modern components are designated as mixed groundwater. In reality, pre-modern groundwater could contain small fractions of modern groundwater, and modern groundwater could contain small fractions of pre-modern groundwater. Previous investigations have used a range of tritium values from 0.3 to 1.0 TU as thresholds for distinguishing pre-1950 from post-1950 groundwater (Michel, 1989; Plummer and others, 1993, p. 260; Michel and Schroeder, 1994; Clark and Fritz, 1997, p. 185; Manning and others, 2005). By using a tritium value of 1.0 TU for the threshold in this study, the age classification scheme allows a larger fraction of modern groundwater to be classified as pre-modern than if a lower threshold was used. A lower threshold for tritium would result in more samples classified as mixed age rather than pre-modern age, when other tracers—^{14}C and terrigenic He— indicated that the samples primarily were pre-modern age. This high threshold was considered more appropriate for this study because many of the wells were production wells with long screens and mixing water of pre-modern and modern age likely occurs.

Geochemical Conditions

Geochemical conditions investigated as potential explanatory variables in this report include oxidation-reduction characteristics, dissolved oxygen concentrations, and pH (tables D7, D8, D9). Oxidation-reduction (redox) conditions and pH influence the mobility of many organic and inorganic constituents (Hem, 1985; McMahon and Chapelle, 2008). Along groundwater flow paths, redox conditions commonly proceed along a well-documented sequence of Terminal Electron Acceptor Processes (TEAP); one TEAP typically dominates at a particular time and aquifer location (Chapelle and others, 1995; Chapelle, 2001). The predominant TEAPs are oxygen-reducing (oxic), nitrate-reducing, manganese-reducing, iron-reducing, sulfate-reducing, and methanogenesis. The presence of redox-sensitive chemical species indicating more than one TEAP may indicate (1) the discharge from the well includes mixed waters from different redox zones upgradient of the well, (2) the well is screened across more than one redox zone, or (3) there is spatial heterogeneity in microbial activity in the aquifer. In addition, different redox couples often are not consistent, indicating electrochemical disequilibrium in groundwater (Lindburg and Runnels, 1984) complicating the assessments of redox conditions.

In this report, redox conditions were represented in two ways: dissolved oxygen concentration and classified redox state. Dissolved oxygen concentrations were measured at USGS-grid and USGS-understanding wells (Burton and Belitz, 2008; Shelton and others, 2008), but are not reported in the CDPH database (tables D7 and D8). Redox conditions were classified based on dissolved oxygen, nitrate, manganese, iron, and sulfate concentrations using the classification scheme of McMahon and Chapelle (2008) (tables D7 and D8). An automated workbook program was used to assign the redox classification to each sample (Jurgens and others, 2009). For wells without USGS inorganic constituent data, the most recent data within the previous 3 years (January 1, 2003, to December 31, 2005) for that well in the CDPH database were used.

Redox conditions for wells in the Domestic Well Project are included in table D9. However, dissolved oxygen data was not collected as part of this project.

Table D1. Nomenclature and well-depth information from domestic wells sampled in Tulare County and located in the Southeast San Joaquin Valley study unit, California GAMA Domestic Well Project, 2006.

[Domestic well is well primarily used for home use. **Abbreviations**: GAMA, Groundwater Ambient Monitoring and Assessment Program; TUL, Tulare County domestic well; ft, feet; LSD, land surface datum; na, not available]

Well identification number	Groundwater subbasin	Well depth (ft below LSD)	Well identification number	Groundwater subbasin	Well depth (ft below LSD)
TUL901	Kaweah	250	TUL960	Tule	na
TUL902	Kaweah	na	TUL961	Tule	320
TUL903	Kaweah	110	TUL967	Kaweah	na
TUL904	Kaweah	130	TUL968	Kings	190
TUL905	Kaweah	80	TUL969	Kings	na
TUL906	Kaweah	na	TUL970	Kings	160
TUL907	Kaweah	na	TUL971	Kings	100
TUL908	Kaweah	na	TUL972	Kings	na
TUL909	Kaweah	98	TUL973	Kings	115
TUL910	Kaweah	na	TUL974	Kaweah	220
TUL911	Kaweah	150	TUL976	Kaweah	179
TUL912	Kaweah	200	TUL978	Tule	380
TUL913	Kaweah	25	TUL979	Tule	na
TUL914	Kaweah	90	TUL986	Tule	na
TUL915	Kaweah	325	TUL987	Tule	na
TUL916	Kaweah	192	TUL988	Kaweah	300
TUL917	Kaweah	150	TUL990	Tule	320
TUL922	Kaweah	200	TUL991	Kaweah	na
TUL923	Kaweah	208	TUL992	Tule	280
TUL924	Kaweah	150	TUL996	Kaweah	160
TUL925	Kaweah	230	TUL997	Kaweah	250
TUL926	Kaweah	100	TUL999	Tule	380
TUL928	Kings	138	TUL1000	Tule	365
TUL929	Kaweah	134	TUL1003	Kaweah	276
TUL930	Kaweah	192	TUL1005	Tule	200
TUL931	Kaweah	na	TUL1006	Tule	na
TUL932	Kaweah	na	TUL1007	Kings	80
TUL933	Kaweah	320	TUL1008	Kaweah	na
TUL934	Kaweah	na	TUL1010	Kings	200
TUL935	Kaweah	130	TUL1011	Tule	300
TUL936	Kaweah	212	TUL1012	Tule	420
TUL937	Kaweah	275	TUL1013	Kaweah	na
TUL938	Kaweah	na	TUL1014	Kings	100
TUL939	Kaweah	189	TUL1015	Kings	90
TUL940	Tule	120	TUL1016	Kings	na
TUL941	Kaweah	312	TUL1017	Kaweah	143
TUL942	Tule	na	TUL1020	Kaweah	na
TUL943	Tule	na	TUL1021	Kings	na
TUL944	Tule	na	TUL1022	Tule	420
TUL945	Tule	320	TUL1025	Kings	120
TUL946	Tule	216	TUL1026	Kaweah	120
TUL947	Tule	na	TUL1027	Kaweah	200
TUL948	Tule	na	TUL1028	Tule	na
TUL949	Kaweah	na	TUL1029	Tule	510
TUL950	Tule	250	TUL1031	Kings	90
TUL951	Tule	na	TUL1032	Kings	200
TUL952	Tule	900	TUL1033	Kaweah	250
TUL955	Kaweah	257	TUL1034	Tule	80
TUL958	Tule	300	TUL1035	Kaweah	154
TUL959	Tule	148	TUL1036	Kaweah	120
			TUL1041	Tule	490

Table D1.   Nomenclature and well-depth information from domestic wells sampled in Tulare County and located in the Southeast San Joaquin Valley study unit, California GAMA Domestic Well Project, 2006.—Continued

[Domestic well is well primarily used for home use. **Abbreviations**: GAMA, Groundwater Ambient Monitoring and Assessment Program; TUL, Tulare County domestic well; ft, feet; LSD, land surface datum; na, not available]

Well identification number	Groundwater subbasin	Well depth (ft below LSD)	Well identification number	Groundwater subbasin	Well depth (ft below LSD)
TUL1042	Kaweah	53	TUL1076	Kaweah	130
TUL1043	Kings	170	TUL1080	Kaweah	130
TUL1044	Kaweah	na	TUL1081	Kaweah	na
TUL1054	Tule	50	TUL1082	Kaweah	na
TUL1055	Kaweah	na	TUL1083	Kaweah	165
TUL1056	Kaweah	300	TUL1084	Kings	na
TUL1057	Kings	na	TUL1085	Kings	100
TUL1058	Kaweah	95	TUL1088	Kaweah	na
TUL1059	Kaweah	152	TUL1089	Kings	na
TUL1060	Kaweah	200	TUL1091	Kaweah	na
TUL1061	Kaweah	300	TUL1092	Kings	na
TUL1062	Tule	68	TUL1093	Kings	100
TUL1064	Tule	160	TUL1094	Kaweah	150
TUL1065	Kaweah	440	TUL1098	Kaweah	na
TUL1066	Kaweah	200	TUL1101	Tule	na
TUL1070	Kings	65	TUL1103	Kaweah	na
TUL1071	Kaweah	315	TUL1105	Kings	165
TUL1072	Kings	75	TUL1106	Kaweah	100
TUL1073	Kings	140	TUL1107	Tule	na
TUL1074	Kings	na	TUL1111	Kings	na
TUL1075	Kaweah	150	TUL1505	Kaweah	163

Table D2.   Land-use classification, normalized lateral position, septic systems, and formerly leaking underground fuel tank information for USGS-grid and USGS-understanding wells, and CDPH-grid wells for inorganic constituents, Southeast San Joaquin Valley study unit, California GAMA Priority Basin Project.

[Land-use classification based on 500-meter radius around each well (Johnson and Belitz, 2009). Number of septic tanks or cesspools in 500-meter radius around each well (U.S. Census Bureau, 1990). Number of leaking underground fuel tanks within a Thiessen polygon in square kilometers, data from Geographic Information Management System GeoTracker (California Environmental Protection Agency, 2001). **Abbreviations:** USGS, U.S. Geological Survey; GAMA, Groundwater Ambient Monitoring and Assessment Program; CDPH, California Department of Public Health; KING, Kings study area well; KWH, Kaweah study area well; TLR, Tulare Lake study area well; TULE, Tule study area well; FP, flow-path well; Hwy99T, transect well; DG, designates well with CDPH data from same well as USGS-grid well; DPH, designates well selected from subset of CDPH wells; km², square kilometer]

USGS GAMA well identification number	CDPH GAMA well identification number	Land use (in percent)				Land-use classification	Normalized lateral position from valley trough	Number of septic tanks or cesspools	Density of formerly leaking underground fuel tanks (number of tanks/km²)
		Orchards or vineyards[1]	Agricultural	Natural	Urban				
Grid wells									
KING-01	KING-DG-01	1.4	75.7	6.8	17.5	Agricultural	0.04	0.7	0.02
KING-02		2.4	89.6	0.0	10.4	Agricultural	0.10	2.3	0.01
KING-03	KING-DG-03	2.1	30.8	0.7	68.5	Urban	0.18	2.4	0.04
KING-04		16.0	88.3	2.1	9.6	Agricultural	0.20	1.3	0.01
KING-05		7.4	91.5	7.2	1.3	Agricultural	0.12	1.1	0.01
KING-06		26.9	30.0	12.4	57.6	Urban	0.35	15.4	0.07
KING-07	KING-DG-07	4.1	36.3	8.8	54.9	Urban	0.36	7.3	0.02
KING-08		49.9	55.0	2.6	42.4	Agricultural	0.76	4.2	0.34
KING-09	KING-DG-09	27.0	55.2	11.1	33.7	Agricultural	0.61	18.3	0.31
KING-10		6.8	10.3	5.4	84.3	Urban	0.60	9.6	0.43
KING-11	KING-DG-11	5.0	5.6	0.9	93.5	Urban	0.83	4.3	0.25
KING-12	KING-DG-12	3.7	7.4	2.3	90.3	Urban	0.85	32.7	1.02
KING-13	KING-DG-13	39.5	60.1	30.6	9.3	Agricultural	0.80	3.0	1.32
KING-14	KING-DG-14	11.8	67.6	0.8	31.6	Agricultural	0.34	6.8	0.21
KING-15	KING-DG-15	0.0	0.0	18.6	81.4	Urban	0.65	15.8	1.00
KING-16	KING-DG-16	0.0	0.0	2.5	97.5	Urban	0.70	0.0	0.65
KING-17		2.3	39.5	16.4	44.1	Mixed	0.63	15.6	0.42
KING-18	KING-DG-18	0.1	0.1	4.7	95.2	Urban	0.86	3.5	0.72
KING-19	KING-DG-19	25.7	33.8	3.8	62.4	Urban	0.66	3.2	0.02
KING-20		19.6	85.5	10.2	4.4	Agricultural	0.96	7.3	0.07
KING-21	KING-DG-21	4.7	98.9	0.0	1.1	Agricultural	0.11	8.2	0.02
KING-22	KING-DG-22	0.0	7.2	47.9	44.9	Mixed	0.83	2.8	0.65
KING-23	KING-DG-23	0.0	12.8	7.9	79.3	Urban	0.89	9.9	0.15
KING-24	KING-DG-24	0.0	1.1	24.2	74.7	Urban	0.78	37.3	1.39
KING-25	KING-DG-25	21.1	44.4	12.1	43.4	Mixed	0.59	2.7	0.10
KING-26		18.4	67.2	32.8	0.0	Agricultural	0.06	1.2	0.03
KING-27		98.7	98.7	0.0	1.3	Agricultural	0.36	6.8	0.03
KING-28		0.0	0.1	91.2	8.7	Natural	1.00	0.7	0.02
KING-29	KING-DG-29	3.2	44.3	54.6	1.0	Natural	0.96	16.7	0.09
KING-30		54.0	83.3	15.3	1.4	Agricultural	0.91	4.7	0.03
KING-31		0.0	88.3	7.0	4.7	Agricultural	0.03	0.9	0.02
KING-32		86.9	89.0	6.6	4.4	Agricultural	0.41	7.7	0.03
KING-33		0.9	65.3	8.1	26.6	Agricultural	0.80	1.5	0.03
KING-34	KING-DG-34	60.1	75.1	7.3	17.5	Agricultural	0.37	5.1	0.03
KING-35	KING-DG-35	60.7	81.0	1.8	17.2	Agricultural	0.52	5.3	0.08
KING-36	KING-DG-36	71.9	74.5	8.2	17.3	Agricultural	0.48	5.8	0.03
KING-37		36.0	96.9	2.7	0.3	Agricultural	0.47	5.7	0.02
KING-38		76.6	99.0	1.0	0.0	Agricultural	0.66	7.1	0.28
KING-39		6.0	17.4	27.4	55.2	Urban	0.09	2.0	0.01
KWH-01	KWH-DG-01	28.8	91.3	6.6	2.1	Agricultural	0.80	35.9	0.36
KWH-02	KWH-DG-02	22.0	47.4	7.0	45.6	Mixed	0.85	18.9	0.05
KWH-03	KWH-DG-03	31.4	65.3	8.9	25.8	Agricultural	0.65	9.7	0.06

Table D2. Land-use classification, normalized lateral position, septic systems, and formerly leaking underground fuel tank information for USGS-grid and USGS-understanding wells, and CDPH-grid wells for inorganic constituents, Southeast San Joaquin Valley study unit, California GAMA Priority Basin Project.—Continued

[Land-use classification based on 500-meter radius around each well (Johnson and Belitz, 2009). Number of septic tanks or cesspools in 500-meter radius around each well (U.S. Census Bureau, 1990). Number of leaking underground fuel tanks within a Thiessen polygon in square kilometers, data from Geographic Information Management System GeoTracker (California Environmental Protection Agency, 2001). **Abbreviations:** USGS, U.S. Geological Survey; GAMA, Groundwater Ambient Monitoring and Assessment Program; CDPH, California Department of Public Health; KING, Kings study area well; KWH, Kaweah study area well; TLR, Tulare Lake study area well; TULE, Tule study area well; FP, flow-path well; Hwy99T, transect well; DG, designates well with CDPH data from same well as USGS-grid well; DPH, designates well selected from subset of CDPH wells; km², square kilometer]

USGS GAMA well identification number	CDPH GAMA well identification number	Land use (in percent)				Land-use classification	Normalized lateral position from valley trough	Number of septic tanks or cesspools	Density of formerly leaking underground fuel tanks (number of tanks/km²)
		Orchards or vineyards[1]	Agricultural	Natural	Urban				
Grid wells—Continued									
KWH-04	KWH-DG-04	28.4	70.7	25.2	4.1	Agricultural	1.00	4.9	0.13
KWH-05	KWH-DG-05	74.1	89.7	8.8	1.5	Agricultural	1.00	3.2	0.04
KWH-06	KWH-DG-06	15.8	44.9	11.7	43.4	Mixed	0.92	8.7	1.04
KWH-07		57.6	73.4	18.8	7.8	Agricultural	1.00	7.5	0.06
KWH-08	KWH-DG-08	49.3	77.0	22.8	0.2	Agricultural	1.00	5.6	0.04
KWH-09		24.2	85.7	3.3	11.0	Agricultural	0.72	7.9	0.05
KWH-10		4.1	6.5	14.8	78.7	Urban	0.98	28.1	1.07
KWH-11	KWH-DG-11	0.1	66.1	9.0	24.9	Agricultural	0.46	9.5	0.32
KWH-12		0.0	0.0	17.0	83.0	Urban	0.53	1.3	3.60
KWH-13	KWH-DG-13	1.6	83.4	16.6	0.0	Agricultural	0.35	2.5	0.05
KWH-14		0.1	14.9	34.8	50.3	Urban	0.56	4.3	0.06
KWH-15	KWH-DG-15	0.0	98.6	1.4	0.0	Agricultural	0.11	2.3	0.06
KWH-16		23.8	89.3	3.6	7.1	Agricultural	0.63	7.3	0.03
KWH-17		16.8	79.7	12.9	7.3	Agricultural	0.18	2.5	0.01
KWH-18		15.7	75.7	4.9	19.4	Agricultural	0.78	19.3	0.03
TLR-01	TLR-DG-01	0.9	77.8	16.0	6.2	Agricultural	0.10	2.4	0.07
TLR-02		8.1	47.7	3.9	48.5	Mixed	0.14	4.9	0.22
TLR-03	TLR-DG-03	3.3	63.0	6.9	30.1	Agricultural	0.18	13.3	0.08
TLR-04		4.4	76.7	21.1	2.2	Agricultural	0.26	4.2	0.04
TLR-05		0.5	99.2	0.8	0.0	Agricultural	0.20	3.1	0.02
TLR-06	TLR-DG-06	2.4	59.9	4.4	35.7	Agricultural	0.00	5.8	0.19
TLR-07	TLR-DG-07	0.0	4.1	48.1	47.8	Mixed	0.18	0.7	0.03
TLR-08		0.2	44.4	5.6	49.9	Mixed	0.26	6.8	0.43
TLR-09		0.0	90.4	2.6	7.0	Agricultural	0.00	0.2	0.00
TULE-01		7.0	16.2	7.0	76.9	Urban	0.44	3.1	0.19
TULE-02	TULE-DG-02	6.9	40.9	3.1	56.0	Urban	0.66	2.8	0.02
TULE-03	TULE-DG-03	1.5	73.2	16.6	10.2	Agricultural	0.42	2.0	0.02
TULE-04	TULE-DG-04	40.9	61.5	12.8	25.7	Agricultural	0.71	0.3	0.03
TULE-05	TULE-DG-05	8.4	64.1	6.3	29.6	Agricultural	0.85	3.3	0.01
TULE-06		69.9	99.9	0.1	0.0	Agricultural	0.80	5.2	0.05
TULE-07	TULE-DG-07	0.0	12.5	6.6	80.9	Urban	0.42	0.1	0.03
TULE-08		0.2	38.9	0.7	60.4	Urban	0.45	6.2	0.14
TULE-09	TULE-DG-09	0.0	5.0	8.6	86.4	Urban	0.98	71.2	3.57
TULE-10		51.7	86.3	13.2	0.6	Agricultural	0.83	5.4	0.05
TULE-11		0.0	100.0	0.0	0.0	Agricultural	0.18	1.1	0.01
TULE-12		0.1	26.8	63.7	9.5	Natural	0.18	1.1	0.01
TULE-13	TULE-DG-13	14.7	39.6	1.5	58.9	Urban	0.76	55.5	0.03
TULE-14	TULE-DG-14	25.4	93.1	6.6	0.2	Agricultural	0.65	1.2	0.01
TULE-15		0.0	99.3	0.6	0.1	Agricultural	0.13	1.1	0.01
TULE-16		86.3	99.3	0.5	0.2	Agricultural	0.55	0.6	0.02
TULE-17	TULE-DG-17	0.0	100.0	0.0	0.0	Agricultural	0.36	0.9	0.01
	KING-DPH-06	37.3	92.0	2.9	5.2	Agricultural	0.33	4.5	0.03

Table D2. Land-use classification, normalized lateral position, septic systems, and formerly leaking underground fuel tank information for USGS-grid and USGS-understanding wells, and CDPH-grid wells for inorganic constituents, Southeast San Joaquin Valley study unit, California GAMA Priority Basin Project.—Continued

[Land-use classification based on 500-meter radius around each well (Johnson and Belitz, 2009). Number of septic tanks or cesspools in 500-meter radius around each well (U.S. Census Bureau, 1990). Number of leaking underground fuel tanks within a Thiessen polygon in square kilometers, data from Geographic Information Management System GeoTracker (California Environmental Protection Agency, 2001). **Abbreviations**: USGS, U.S. Geological Survey; GAMA, Groundwater Ambient Monitoring and Assessment Program; CDPH, California Department of Public Health; KING, Kings study area well; KWH, Kaweah study area well; TLR, Tulare Lake study area well; TULE, Tule study area well; FP, flow-path well; Hwy99T, transect well; DG, designates well with CDPH data from same well as USGS-grid well; DPH, designates well selected from subset of CDPH wells; km², square kilometer]

USGS GAMA well identification number	CDPH GAMA well identification number	Land use (in percent)				Land-use classification	Normalized lateral position from valley trough	Number of septic tanks or cesspools	Density of formerly leaking underground fuel tanks (number of tanks/km²)
		Orchards or vineyards[1]	Agricultural	Natural	Urban				
Grid wells—Continued									
	KING-DPH-08	76.7	83.2	6.6	10.2	Agricultural	0.66	12.7	0.06
	KING-DPH-10	53.7	83.7	11.5	4.8	Agricultural	0.66	10.0	0.13
	KING-DPH-14	12.5	77.3	8.9	13.7	Agricultural	0.08	1.4	0.01
	KING-DPH-27	67.2	68.2	0.7	31.2	Agricultural	0.50	4.5	0.03
	KING-DPH-28	0.0	0.0	100.0	0.0	Natural	1.00	0.7	0.02
	KING-DPH-32	95.2	96.0	1.0	3.0	Agricultural	0.47	10.5	0.06
	KING-DPH-38	20.5	26.7	19.2	54.1	Urban	0.65	8.1	9.23
	KING-DPH-39	25.0	76.7	0.7	22.6	Agricultural	0.30	2.8	0.19
	KING-DPH-40	0.0	64.9	4.8	30.2	Agricultural	0.04	0.9	0.02
	KWH-DPH-07	33.1	39.3	39.3	21.4	Mixed	1.00	6.4	0.01
	KWH-DPH-08	30.7	62.0	20.0	18.0	Agricultural	0.94	8.9	0.27
	KWH-DPH-09	34.4	89.0	9.6	1.4	Agricultural	0.82	4.6	0.03
	KWH-DPH-10	7.4	66.2	7.4	26.3	Agricultural	0.83	9.8	0.03
	KWH-DPH-17	7.8	82.2	17.8	0.0	Agricultural	0.12	0.8	0.01
	KWH-DPH-19	71.8	97.5	1.7	0.8	Agricultural	0.37	2.5	0.03
	KWH-DPH-20	11.2	93.8	3.2	3.0	Agricultural	0.42	2.7	0.03
	KWH-DPH-21	26.7	58.2	3.8	38.0	Agricultural	0.61	7.3	0.03
	TLR-DPH-08	1.6	33.7	5.5	60.8	Urban	0.26	6.8	1.62
	TLR-DPH-09	2.3	86.5	9.3	4.2	Agricultural	0.00	0.1	0.00
	TULE-DPH-06	4.9	58.6	4.2	37.1	Agricultural	0.88	27.8	0.45
	TULE-DPH-10	0.0	4.1	95.9	0.0	Natural	1.00	0.2	0.05
	TULE-DPH-16	37.5	94.4	3.9	1.7	Agricultural	0.54	0.8	0.02
	TULE-DPH-18	0.0	93.1	6.9	0.0	Agricultural	0.06	0.9	0.01
Understanding wells									
KINGFP-01		21.3	80.5	14.7	4.8	Agricultural	0.72	17.4	0.29
KINGFP-02		22.7	79.8	15.0	5.2	Agricultural	0.72	17.4	0.29
KINGFP-03		5.2	23.4	12.6	64.0	Urban	0.70	13.7	0.80
KINGFP-04		0.0	2.5	23.5	74.0	Urban	0.70	9.1	1.43
KINGFP-05		91.1	95.4	3.0	1.6	Agricultural	0.81	6.1	0.06
KINGFP-06		91.1	95.4	3.0	1.6	Agricultural	0.81	6.1	0.06
KINGFP-07		91.1	95.4	3.0	1.6	Agricultural	0.81	6.1	0.06
KINGFP-08		72.2	80.0	3.0	17.1	Agricultural	0.85	7.7	0.06
KINGFP-09		72.2	80.0	3.0	17.1	Agricultural	0.85	7.7	0.06
KINGFP-10		72.2	80.0	3.0	17.1	Agricultural	0.85	7.7	0.06
KINGFP-11		75.5	90.1	5.3	4.6	Agricultural	0.79	9.6	0.14
KINGFP-12		75.5	90.1	5.3	4.6	Agricultural	0.79	9.6	0.14
KINGFP-13		2.2	40.5	16.5	43.0	Mixed	0.63	15.9	0.42
KINGFP-14		46.7	48.6	36.0	15.5	Mixed	0.47	5.1	0.02
KINGFP-15		46.7	48.6	36.0	15.5	Mixed	0.47	5.1	0.02
HWY99T-01		4.9	45.7	19.2	35.1	Mixed	0.67	6.4	0.12

[1] Orchard and vineyard land use is a subset of agricultural land use.

Table D3. Land-use classification, normalized lateral position, septic systems, and formerly leaking underground fuel tank information for USGS-grid and USGS-understanding wells, and CDPH-grid wells for inorganic constituents, Kern County Subbasin study unit, California GAMA Priority Basin Project.

[Land-use classification based on 500-meter radius around each well (Johnson and Belitz, 2009). Number of septic tanks or cesspools in 500-meter radius around each well (U.S. Census Bureau, 1990). Number of leaking underground fuel tanks within a Thiessen polygon in square kilometers, data from Geographic Information Management System GeoTracker (California Environmental Protection Agency, 2001). **Abbreviations**: USGS, U.S. Geological Survey; GAMA, Groundwater Ambient Monitoring and Assessment Program; CDPH, California Department of Public Health; km^2, square kilometer; KERN, Kern study area well; FP, flow-path well; DG, designates well with CDPH data from same well as USGS-grid well; DPH, designates well selected from subset of CDPH wells; na, not available]

USGS GAMA well identification number	CDPH GAMA well identification number	Land use (in percent)				Land-use classification	Normalized lateral position from valley trough	Number of septic tanks or cesspools	Density of formerly leaking underground fuel tanks (number of tanks/km^2)
		Orchards or vineyards	Agricultural	Natural	Urban				
Grid wells									
KERN-01		0.0	5.6	94.3	0.1	Natural	0.99	0.61	0.004
KERN-02		41.9	49.9	13.2	36.9	Mixed	0.73	5.76	0.02
KERN-03	KERN-DG-03	3.4	47.1	2.5	50.4	Urban	0.40	1.79	0.08
KERN-04	KERN-DG-04	0.0	0.0	11.3	88.7	Urban	0.59	13.18	4.17
KERN-05		4.2	91.2	7.8	1.0	Agricultural	0.00	0.36	0.004
KERN-06	KERN-DG-06	0.8	100.0	0.0	0.0	Agricultural	0.31	0.13	0.01
KERN-07	KERN-DG-07	9.6	64.5	8.4	27.1	Agricultural	0.51	2.75	0.56
KERN-08	KERN-DG-08	21.5	71.8	11.8	16.4	Agricultural	0.08	0.74	0.04
KERN-09	KERN-DG-09	0.0	53.3	46.6	0.1	Agricultural	0.04	0.74	0.02
KERN-10	KERN-DG-10	0.0	93.4	4.0	2.6	Agricultural	0.00	0.32	0.004
KERN-11	KERN-DG-11	45.1	71.7	14.2	14.1	Agricultural	0.72	0.39	0.02
KERN-12	KERN-DG-12	2.7	96.2	3.8	0.0	Agricultural	0.01	0.74	0.004
KERN-13	KERN-DG-13	0.2	71.2	2.3	26.5	Agricultural	0.31	1.42	0.01
KERN-14	KERN-DG-14	28.5	45.7	29.6	24.7	Mixed	0.68	0.47	0.01
KERN-15	KERN-DG-15	0.5	59.5	10.4	30.1	Agricultural	0.34	76.27	0.88
KERN-16		0.3	96.0	4.0	0.0	Agricultural	0.00	0.74	0.01
KERN-17	KERN-DG-17	0.0	11.7	15.7	72.6	Urban	0.48	28.94	0.69
KERN-18	KERN-DG-18	0.2	86.1	13.2	0.7	Agricultural	0.40	0.60	0.12
KERN-19	KERN-DG-19	0.1	84.2	6.9	8.9	Agricultural	0.61	83.72	0.25
KERN-20		0.0	13.9	77.8	8.4	Natural	0.00	0.32	0.01
KERN-21		1.1	97.7	2.3	0.0	Agricultural	0.14	1.61	0.16
KERN-22	KERN-DG-22	1.5	9.5	49.6	40.9	Mixed	0.77	2.41	0.11
KERN-23	KERN-DG-23	0.0	0.1	98.3	1.6	Natural	0.36	0.32	0.004
KERN-24		0.2	93.1	2.7	4.1	Agricultural	0.22	0.13	0.01
KERN-25		32.4	67.7	9.7	22.6	Agricultural	0.63	1.01	0.03
KERN-26	KERN-DG-26	0.9	91.5	8.5	0.0	Agricultural	0.14	1.13	0.01
KERN-27		24.1	93.5	5.6	0.9	Agricultural	0.36	1.61	0.01
KERN-28	KERN-DG-28	20.5	75.6	6.4	18.0	Agricultural	0.60	0.76	0.11
KERN-29	KERN-DG-29	49.5	55.9	44.1	0.0	Agricultural	0.65	0.60	0.01
KERN-30	KERN-DG-30	69.1	93.0	4.5	2.5	Agricultural	0.61	0.39	0.08
KERN-31	KERN-DG-31	0.0	3.2	0.8	96.0	Urban	0.46	218.07	0.82
KERN-32	KERN-DG-32	0.1	96.8	1.0	2.2	Agricultural	0.43	0.60	0.01
KERN-33		16.7	84.8	11.1	4.1	Agricultural	0.49	22.87	0.37
KERN-34		0.0	41.4	58.6	0.0	Natural	0.06	0.74	0.01
KERN-35		0.1	23.9	76.1	0.0	Natural	0.33	1.13	0.03
KERN-36		0.0	41.2	58.8	0.0	Natural	0.53	36.86	0.09
KERN-37		0.0	0.6	99.2	0.2	Natural	0.95	1.74	0.02
KERN-38		0.6	99.4	0.6	0.0	Agricultural	0.91	0.31	0.01
KERN-39		1.4	84.1	12.6	3.3	Agricultural	0.45	1.14	0.05
KERN-40		19.9	54.6	23.3	22.1	Agricultural	0.77	0.42	0.01
KERN-41		3.8	99.3	0.0	0.7	Agricultural	0.44	1.68	0.02
KERN-42		0.0	16.7	40.4	42.8	Mixed	0.38	0.00	0.18
KERN-43	KERN-DG-43	1.0	69.8	9.7	20.5	Agricultural	0.39	0.39	0.11
KERN-44		1.3	44.0	4.8	51.2	Urban	0.43	10.68	0.07

Table D3. Land-use classification, normalized lateral position, septic systems, and formerly leaking underground fuel tank information for USGS-grid and USGS-understanding wells, and CDPH-grid wells for inorganic constituents, Kern County Subbasin study unit, California GAMA Priority Basin Project.—Continued

[Land-use classification based on 500-meter radius around each well (Johnson and Belitz, 2009). Number of septic tanks or cesspools in 500-meter radius around each well (U.S. Census Bureau, 1990). Number of leaking underground fuel tanks within a Thiessen polygon in square kilometers, data from Geographic Information Management System GeoTracker (California Environmental Protection Agency, 2001). **Abbreviations:** USGS, U.S. Geological Survey; GAMA, Groundwater Ambient Monitoring and Assessment Program; CDPH, California Department of Public Health; km², square kilometer; KERN, Kern study area well; FP, flow-path well; DG, designates well with CDPH data from same well as USGS-grid well; DPH, designates well selected from subset of CDPH wells; na, not available]

USGS GAMA well identification number	CDPH GAMA well identification number	Land use (in percent)				Land-use classification	Normalized lateral position from valley trough	Number of septic tanks or cesspools	Density of formerly leaking underground fuel tanks (number of tanks/km²)
		Orchards or vineyards	Agricultural	Natural	Urban				
Grid wells—Continued									
KERN-45		12.4	84.0	14.3	1.7	Agricultural	0.36	11.46	0.27
KERN-46		37.9	88.8	10.2	1.0	Agricultural	0.34	0.59	0.01
KERN-47	KERN-DG-47	18.3	75.6	5.2	19.2	Agricultural	0.48	2.31	0.02
	KERN-DPH-01	56.0	56.0	43.4	0.6	Agricultural	0.67	0.61	0.004
	KERN-DPH-02	71.9	83.3	8.9	7.8	Agricultural	0.70	5.77	0.02
	KERN-DPH-05	2.6	98.1	1.8	0.1	Agricultural	0.00	0.39	0.004
	KERN-DPH-16	0.2	97.6	2.4	0.0	Agricultural	0.00	0.74	na
	KERN-DPH-20	0.2	64.0	33.2	2.7	Agricultural	0.00	0.44	0.01
	KERN-DPH-21	0.9	40.0	6.8	53.3	Urban	0.00	0.91	2.60
	KERN-DPH-24	26.7	98.6	0.6	0.8	Agricultural	0.17	0.13	0.01
	KERN-DPH-25	68.6	96.0	4.0	0.0	Agricultural	0.55	1.01	0.03
	KERN-DPH-27	56.2	90.5	8.7	0.8	Agricultural	0.34	1.61	0.01
	KERN-DPH-28	41.2	84.8	15.2	0.0	Agricultural	0.53	0.76	0.04
	KERN-DPH-30	5.7	66.8	6.4	26.8	Agricultural	0.58	2.88	0.07
	KERN-DPH-31	0.0	0.0	2.6	97.4	Urban	0.54	0.91	1.66
	KERN-DPH-33	10.2	46.5	14.4	39.1	Mixed	0.54	23.75	0.39
	KERN-DPH-46	41.1	87.2	8.2	4.6	Agricultural	0.19	0.89	0.01
	KERN-DPH-48	0.0	93.2	6.0	0.8	Agricultural	0.18	0.13	0.01
	KERN-DPH-49	0.5	21.1	56.8	22.1	Natural	0.11	0.74	0.06
	KERN-DPH-50	3.4	61.9	3.1	35.1	Agricultural	0.42	1.22	2.06
	KERN-DPH-51	38.3	75.3	6.4	18.3	Agricultural	0.45	1.14	0.01
	KERN-DPH-52	0.0	3.8	96.2	0.0	Natural	0.21	0.74	0.02
	KERN-DPH-53	0.0	21.4	78.6	0.0	Natural	0.05	0.74	0.01
	KERN-DPH-54	0.0	80.9	10.2	8.9	Agricultural	0.05	0.32	0.004
	KERN-DPH-55	8.1	83.5	5.7	10.8	Agricultural	0.36	11.26	0.03
	KERN-DPH-56	36.1	60.1	13.1	26.8	Agricultural	0.43	10.01	0.37
	KERN-DPH-57	16.7	68.4	7.0	24.6	Agricultural	0.61	0.39	0.02
	KERN-DPH-58	5.8	42.5	7.0	50.5	Urban	0.58	7.70	0.21
	KERN-DPH-59	13.2	26.2	1.9	71.8	Urban	0.60	3.00	0.05
	KERN-DPH-60	0.0	80.5	7.4	12.0	Agricultural	0.24	0.63	0.004
	KERN-DPH-61	0.3	94.7	1.3	4.0	Agricultural	0.10	1.13	0.01
	KERN-DPH-62	1.9	94.8	2.6	2.5	Agricultural	0.11	0.88	0.02
	KERN-DPH-63	18.9	54.0	43.9	2.2	Agricultural	0.36	0.59	0.02
	KERN-DPH-64	20.0	72.7	26.8	0.5	Agricultural	0.87	0.31	0.01
	KERN-DPH-65	48.5	94.4	1.8	3.8	Agricultural	0.90	0.31	0.01
	KERN-DPH-66	90.4	92.2	0.0	7.8	Agricultural	0.54	0.59	0.01
	KERN-DPH-67	3.7	77.9	7.4	14.7	Agricultural	0.68	1.01	0.02
	KERN-DPH-68	0.7	96.0	1.1	2.9	Agricultural	0.65	11.37	0.11
Understanding wells									
KERNFP-01		0.0	0.0	6.0	94.0	Urban	0.79	1.33	1.15
KERNFP-02		0.0	0.6	77.0	22.4	Natural	0.69	9.77	1.91
KERNFP-03		0.0	0.0	30.7	69.3	Urban	0.60	34.18	0.56

[1] Orchard and vineyard land use is a subset of agricultural land use.

Table D4.   Land-use classification, normalized lateral position, and septic system information for domestic wells sampled in Tulare County and located in the Southeast San Joaquin Valley study unit, California GAMA Domestic Well Project, 2006.

[Land-use classification based on 500-meter radius around each well (Johnson and Belitz, 2009). Number of septic tanks or cesspools in 500-meter radius around each well (U.S. Census Bureau, 1990). **Abbreviations**: GAMA, Groundwater Ambient Monitoring and Assessment Progam; TUL, Tulare County well]

Domestic well identification number	Land use (in percent)				Land-use classification	Normalized lateral position from valley trough	Number of septic tanks or cesspools
	Orchards or vineyards[1]	Agricultural	Natural	Urban			
TUL901	63.1	92.2	7.3	0.5	Agricultural	0.77	12.2
TUL902	30.0	93.7	5.7	0.6	Agricultural	0.81	8.0
TUL903	26.8	82.9	14.3	2.9	Agricultural	0.91	10.2
TUL904	87.0	95.4	4.6	0.0	Agricultural	0.98	3.4
TUL905	54.9	81.5	18.5	0.0	Agricultural	0.94	3.8
TUL906	40.0	84.7	9.8	5.5	Agricultural	0.91	14.4
TUL907	87.1	94.9	4.5	0.7	Agricultural	0.95	5.4
TUL908	59.1	91.9	7.9	0.2	Agricultural	0.97	5.5
TUL909	67.8	88.5	10.5	1.0	Agricultural	0.98	7.5
TUL910	39.4	76.3	23.7	0.0	Agricultural	1.00	7.2
TUL911	34.7	93.3	6.7	0.0	Agricultural	1.00	4.6
TUL912	62.4	86.6	13.4	0.0	Agricultural	1.00	7.2
TUL913	28.8	43.2	49.3	7.5	Mixed	1.00	5.7
TUL914	75.6	79.1	16.8	4.1	Agricultural	1.00	4.0
TUL915	1.8	6.5	93.3	0.2	Natural	1.00	1.0
TUL916	0.0	0.0	99.9	0.1	Natural	1.00	1.0
TUL917	75.8	79.7	20.3	0.0	Agricultural	1.00	7.5
TUL922	89.5	95.9	4.1	0.0	Agricultural	1.00	7.5
TUL923	34.2	60.5	39.2	0.3	Agricultural	1.00	0.9
TUL924	34.0	49.8	23.9	26.4	Mixed	1.00	5.8
TUL925	51.1	95.4	3.8	0.8	Agricultural	1.00	4.9
TUL926	41.1	77.6	7.2	15.2	Agricultural	0.81	10.6
TUL928	22.5	85.7	13.9	0.3	Agricultural	0.99	3.2
TUL929	60.4	94.9	5.1	0.0	Agricultural	0.73	6.1
TUL930	8.9	90.4	6.4	3.2	Agricultural	0.70	6.4
TUL931	56.7	88.7	11.2	0.1	Agricultural	0.93	8.7
TUL932	82.2	89.5	7.5	3.0	Agricultural	0.94	8.8
TUL933	35.4	80.6	19.4	0.0	Agricultural	1.00	0.9
TUL934	31.7	95.0	4.8	0.2	Agricultural	0.79	4.8
TUL935	80.9	90.0	10.0	0.0	Agricultural	0.93	6.1
TUL936	0.6	90.5	9.5	0.0	Agricultural	0.36	2.6
TUL937	15.2	96.7	3.2	0.1	Agricultural	0.30	2.5
TUL938	16.4	77.7	14.6	7.8	Agricultural	0.49	12.1
TUL939	13.6	71.0	22.1	6.8	Agricultural	0.49	15.0
TUL940	61.6	88.2	11.8	0.0	Agricultural	0.81	8.5
TUL941	12.5	55.4	3.4	41.1	Agricultural	0.61	7.3
TUL942	40.5	93.9	4.3	1.7	Agricultural	0.70	3.0
TUL943	49.7	92.2	6.3	1.5	Agricultural	0.83	7.4
TUL944	0.0	0.0	9.4	90.6	Urban	0.99	59.6
TUL945	11.6	64.8	28.8	6.4	Agricultural	1.00	23.6
TUL946	1.5	1.8	9.2	88.9	Urban	0.98	100.8
TUL947	23.0	85.6	13.3	1.1	Agricultural	0.46	1.1
TUL948	29.5	75.3	24.5	0.2	Agricultural	0.49	1.2
TUL949	11.9	93.6	5.8	0.6	Agricultural	0.40	9.5
TUL950	72.8	99.2	0.8	0.0	Agricultural	0.47	0.4
TUL951	83.2	100.0	0.0	0.0	Agricultural	0.50	0.9
TUL952	16.2	99.1	0.9	0.0	Agricultural	0.80	0.6
TUL955	50.7	93.2	6.8	0.0	Agricultural	1.00	2.2
TUL958	50.7	93.2	6.8	0.0	Agricultural	1.00	2.2

Table D4. Land-use classification, normalized lateral position, and septic system information for domestic wells sampled in Tulare County and located in the Southeast San Joaquin Valley study unit, California GAMA Domestic Well Project, 2006.—Continued

[Land-use classification based on 500-meter radius around each well (Johnson and Belitz, 2009). Number of septic tanks or cesspools in 500-meter radius around each well (U.S. Census Bureau, 1990). **Abbreviations**: GAMA, Groundwater Ambient Monitoring and Assessment Progam; TUL, Tulare County well]

Domestic well identification number	Land use (in percent)				Land-use classification	Normalized lateral position from valley trough	Number of septic tanks or cesspools
	Orchards or vineyards[1]	Agricultural	Natural	Urban			
TUL959	33.2	93.3	3.1	3.7	Agricultural	0.96	12.1
TUL960	34.1	93.8	2.4	3.8	Agricultural	0.92	5.1
TUL961	1.4	75.0	25.0	0.0	Agricultural	1.00	1.0
TUL967	19.2	69.7	10.3	20.0	Agricultural	0.79	12.5
TUL968	15.5	34.7	65.3	0.0	Natural	1.00	0.8
TUL969	1.1	88.0	10.6	1.4	Agricultural	0.77	1.9
TUL970	71.2	92.1	7.8	0.1	Agricultural	1.00	8.6
TUL971	68.9	87.8	12.0	0.2	Agricultural	1.00	8.6
TUL972	61.2	85.7	11.4	2.9	Agricultural	0.91	4.7
TUL973	53.5	88.2	10.4	1.4	Agricultural	0.90	5.7
TUL974	49.1	73.4	26.6	0.0	Agricultural	1.00	0.9
TUL976	39.8	97.0	2.9	0.1	Agricultural	0.76	6.9
TUL978	1.3	6.6	93.2	0.2	Natural	1.00	0.2
TUL979	43.6	82.2	17.8	0.0	Agricultural	1.00	2.2
TUL986	1.0	99.4	0.0	0.6	Agricultural	0.56	1.2
TUL987	5.6	100.0	0.0	0.0	Agricultural	0.40	0.9
TUL988	10.7	87.2	8.3	4.5	Agricultural	0.79	5.6
TUL990	7.8	22.0	76.6	1.4	Natural	1.00	2.9
TUL991	71.9	93.0	6.7	0.2	Agricultural	1.00	5.9
TUL992	1.7	93.8	6.2	0.0	Agricultural	0.57	2.6
TUL996	2.2	21.3	30.7	47.9	Mixed	0.61	3.7
TUL997	70.9	92.5	7.5	0.0	Agricultural	1.00	8.9
TUL999	1.4	84.3	15.1	0.6	Agricultural	0.45	1.2
TUL1000	24.3	90.9	1.5	7.7	Agricultural	0.43	4.4
TUL1003	44.1	88.2	4.1	7.7	Agricultural	0.66	6.3
TUL1005	25.3	98.1	1.9	0.0	Agricultural	0.79	7.4
TUL1006	31.2	83.4	3.7	12.9	Agricultural	0.94	11.6
TUL1007	75.8	89.2	10.8	0.0	Agricultural	0.73	9.8
TUL1008	62.7	87.6	12.3	0.1	Agricultural	1.00	7.5
TUL1010	67.8	78.2	17.9	3.9	Agricultural	0.65	9.8
TUL1011	0.2	86.4	13.5	0.1	Agricultural	0.45	1.2
TUL1012	4.0	99.9	0.0	0.1	Agricultural	0.49	1.2
TUL1013	1.5	98.3	1.7	0.0	Agricultural	0.48	4.4
TUL1014	74.5	89.1	1.1	9.8	Agricultural	0.71	10.7
TUL1015	45.1	68.9	31.1	0.0	Agricultural	1.00	0.9
TUL1016	60.8	68.8	6.7	24.5	Agricultural	0.98	9.9
TUL1017	6.5	96.7	2.5	0.8	Agricultural	0.42	9.5
TUL1020	44.4	96.2	3.8	0.0	Agricultural	0.67	3.4
TUL1021	58.6	93.6	6.1	0.3	Agricultural	1.00	0.9
TUL1022	1.0	95.7	2.9	1.5	Agricultural	0.00	0.8
TUL1025	72.1	82.7	11.7	5.6	Agricultural	0.88	6.7
TUL1026	30.0	57.1	42.9	0.0	Agricultural	1.00	0.9
TUL1027	35.0	86.4	10.5	3.1	Agricultural	0.76	15.4
TUL1028	4.0	99.7	0.3	0.0	Agricultural	0.48	1.3
TUL1029	1.3	86.2	13.8	0.0	Agricultural	0.59	2.0
TUL1031	46.0	70.4	26.3	3.3	Agricultural	0.66	9.8
TUL1032	84.5	94.6	5.3	0.1	Agricultural	0.83	6.8
TUL1033	0.5	65.0	4.7	30.3	Agricultural	0.48	4.4

Table D4. Land-use classification, normalized lateral position, and septic system information for domestic wells sampled in Tulare County and located in the Southeast San Joaquin Valley study unit, California GAMA Domestic Well Project, 2006.—Continued

[Land-use classification based on 500-meter radius around each well (Johnson and Belitz, 2009). Number of septic tanks or cesspools in 500-meter radius around each well (U.S. Census Bureau, 1990). **Abbreviations**: GAMA, Groundwater Ambient Monitoring and Assessment Progam; TUL, Tulare County well]

Domestic well identification number	Land use (in percent)				Land-use classification	Normalized lateral position from valley trough	Number of septic tanks or cesspools
	Orchards or vineyards[1]	Agricultural	Natural	Urban			
TUL1034	63.2	92.4	7.4	0.2	Agricultural	0.90	4.8
TUL1035	45.4	87.5	12.3	0.1	Agricultural	0.72	10.6
TUL1036	1.1	80.5	19.5	0.0	Agricultural	0.67	8.9
TUL1041	4.6	99.8	0.0	0.2	Agricultural	0.50	1.2
TUL1042	52.9	91.9	8.0	0.1	Agricultural	0.91	10.2
TUL1043	76.4	95.9	4.1	0.0	Agricultural	0.76	8.4
TUL1044	36.2	80.0	20.0	0.0	Agricultural	1.00	7.2
TUL1054	0.0	99.3	0.6	0.1	Agricultural	0.34	0.5
TUL1055	10.1	69.1	11.2	19.7	Agricultural	0.31	2.6
TUL1056	24.2	64.0	5.3	30.7	Agricultural	0.87	10.1
TUL1057	60.5	71.6	17.1	11.3	Agricultural	0.85	13.1
TUL1058	38.6	56.4	17.6	26.0	Agricultural	1.00	7.5
TUL1059	42.1	59.0	17.6	23.4	Agricultural	1.00	7.5
TUL1060	75.5	93.3	6.7	0.0	Agricultural	0.91	8.1
TUL1061	54.8	72.9	20.2	6.8	Agricultural	1.00	7.5
TUL1062	0.0	15.8	14.2	70.1	Urban	1.00	25.0
TUL1064	6.3	91.7	8.0	0.3	Agricultural	0.50	2.6
TUL1065	1.1	95.8	4.1	0.1	Agricultural	0.23	2.5
TUL1066	11.2	73.3	11.1	15.6	Agricultural	0.31	2.6
TUL1070	68.5	79.8	19.7	0.6	Agricultural	0.82	6.7
TUL1071	9.6	96.0	4.0	0.0	Agricultural	0.36	2.5
TUL1072	21.2	81.0	17.4	1.6	Agricultural	0.81	1.5
TUL1073	84.2	90.9	8.7	0.5	Agricultural	0.87	6.7
TUL1074	82.3	92.2	7.7	0.1	Agricultural	0.87	6.7
TUL1075	64.3	89.2	10.8	0.0	Agricultural	0.94	9.5
TUL1076	47.4	83.2	11.2	5.6	Agricultural	0.90	7.2
TUL1080	9.9	97.4	2.1	0.6	Agricultural	0.81	12.1
TUL1081	72.1	93.3	5.9	0.8	Agricultural	0.73	6.1
TUL1082	27.7	93.9	4.2	1.8	Agricultural	0.69	6.1
TUL1083	71.1	94.3	5.5	0.2	Agricultural	0.76	12.2
TUL1084	83.9	92.8	2.9	4.3	Agricultural	0.62	9.7
TUL1085	15.5	34.7	65.3	0.0	Natural	1.00	0.8
TUL1088	9.1	65.8	34.2	0.0	Agricultural	1.00	0.9
TUL1089	87.8	96.2	3.8	0.0	Agricultural	1.00	8.4
TUL1091	0.0	0.0	99.9	0.1	Natural	1.00	1.0
TUL1092	70.4	96.8	3.2	0.0	Agricultural	0.72	2.0
TUL1093	4.7	97.6	2.4	0.0	Agricultural	0.79	1.5
TUL1094	4.3	40.3	58.4	1.3	Natural	1.00	0.9
TUL1098	3.3	71.4	7.2	21.4	Agricultural	0.68	3.4
TUL1101	58.4	77.9	22.1	0.0	Agricultural	1.00	2.2
TUL1103	36.8	83.4	14.3	2.3	Agricultural	0.80	18.8
TUL1105	61.6	93.5	6.5	0.0	Agricultural	0.88	5.9
TUL1106	20.8	92.7	6.9	0.5	Agricultural	0.78	18.9
TUL1107	46.1	83.0	15.3	1.7	Agricultural	1.00	2.2
TUL1111	56.2	86.9	13.1	0.0	Agricultural	1.00	8.6
TUL1505	56.6	92.0	8.0	0.0	Agricultural	0.93	3.2

[1] Orchard and vineyard land use is a subset of agricultural land use.

Table D5. Groundwater age classification information for USGS-grid and USGS-understanding wells sampled October 2005 through February 2006, Southeast San Joaquin Valley study unit, California GAMA Priority Basin Project.

[Samples classified as pre-modern if recharged berfore 1952. Samples classified as modern if recharged after 1952. Samples classified as mixed if sample contains both modern and pre-modern water. **Abbreviations**: USGS, U.S. Geological Survey; GAMA, Groundwater Ambient Monitoring and Assessment Program; °C, degrees Celsius; TU, tritium units; KING, Kings study area well; KWH, Kaweah study area well; TLR, Tulare Lake study area well; TULE, Tule study area well; FP, flow-path well; HWY99T, transect well; na, not available; <, less than; nc, not collected]

USGS GAMA well identification number	Noble-gas based recharge temperature (°C)	Tritium activity (TU)	Terrigenic helium, (percent of total helium)	Modern carbon-14 (percent)	Groundwater age classification	USGS GAMA well identification number	Noble-gas based recharge temperature (°C)	Tritium activity (TU)	Terrigenic helium, (percent of total helium)	Modern carbon-14 (percent)	Groundwater age classification
Grid wells						**Grid wells—Continued**					
KING-01	14	<1	82.1	nc	Pre-Modern	KWH-04	17	4.4	75.7	nc	Mixed
KING-02	18	1.2	92.3	nc	Mixed	KWH-05	20	7.6	0.0	nc	Modern
KING-03	na	4.9	88.5	nc	Mixed	KWH-06	20	<1	74.8	nc	Pre-Modern
KING-04	20	2.4	82.6	60.5	Mixed	KWH-07	20	5.0	94.4	nc	Mixed
KING-05	15	<1	89.9	nc	Pre-Modern	KWH-08	20	3.1	93.2	nc	Mixed
KING-06	16	<1	84.0	nc	Pre-Modern	KWH-09	18	9.4	0.0	nc	Modern
KING-07	17	7.4	0.0	nc	Modern	KWH-10	21	3.2	52.5	nc	Mixed
KING-08	15	<1	0.0	nc	Mixed	KWH-11	15	<1	92.1	nc	Pre-Modern
KING-09	16	1.6	0.0	nc	Modern	KWH-12	17	5.0	0.0	85.7	Mixed
KING-10	17	5.0	0.0	nc	Modern	KWH-13	19	5.2	0.0	nc	Modern
KING-11	22	<1	72.3	nc	Pre-Modern	KWH-14	15	4.3	0.0	105	Modern
KING-12	20	3.4	0.0	nc	Modern	KWH-15	17	8.9	43.2	nc	Mixed
KING-13	21	<1	64.4	nc	Pre-Modern	KWH-16	18	8.9	0.0	nc	Modern
KING-14	18	<1	55.3	nc	Pre-Modern	KWH-17	21	11.0	0.0	nc	Modern
KING-15	20	<1	30.4	nc	Pre-Modern	KWH-18	19	5.9	3.9	nc	Modern
KING-16	18	1.3	0.4	nc	Modern	TLR-01	nc	nc	nc	nc	Not Datable
KING-17	20	<1	54.3	66.9	Pre-Modern	TLR-02	16	<1	85.6	nc	Pre-Modern
KING-18	15	7.6	0.1	nc	Modern	TLR-03	15	3.5	96.9	nc	Mixed
KING-19	20	1.1	28.2	nc	Mixed	TLR-04	17	<1	98.9	0.73	Pre-Modern
KING-20	15	4.5	0.0	116	Modern	TLR-05	15	<1	94.0	7.66	Pre-Modern
KING-21	17	1.7	68.6	nc	Mixed	TLR-06	14	<1	75.0	nc	Pre-Modern
KING-22	21	4.5	0.0	nc	Modern	TLR-07	14	<1	34.4	nc	Pre-Modern
KING-23	20	8.1	0.0	nc	Modern	TLR-08	17	<1	82.2	nc	Pre-Modern
KING-24	18	6.6	0.0	nc	Modern	TLR-09	na	<1	89.1	nc	Pre-Modern
KING-25	21	<1	77.0	nc	Pre-Modern	TULE-01	17	5.4	4.2	42.5	Modern
KING-26	16	<1	94.2	nc	Pre-Modern	TULE-02	18	7.3	0.0	nc	Modern
KING-27	25	<1	42.0	nc	Pre-Modern	TULE-03	16	6.3	0.0	nc	Modern
KING-28	16	3.7	0.0	nc	Modern	TULE-04	20	<1	25.4	nc	Pre-Modern
KING-29	19	4.3	77.1	nc	Mixed	TULE-05	18	<1	98.9	nc	Pre-Modern
KING-30	19	<1	77.8	nc	Pre-Modern	TULE-06	15	<1	96.5	nc	Pre-Modern
KING-31	40	<1	0.0	nc	Mixed	TULE-07	16	3.7	4.7	nc	Modern
KING-32	15	<1	9.0	nc	Pre-Modern	TULE-08	17	<1	21.6	42.1	Pre-Modern
KING-33	18	<1	99.1	nc	Pre-Modern	TULE-09	16	3.6	0.0	nc	Modern
KING-34	22	9.8	4.9	nc	Modern	TULE-10	19	1.9	0.0	nc	Modern
KING-35	21	2.8	3.1	nc	Modern	TULE-11	13	<1	49.5	nc	Pre-Modern
KING-36	17	<1	0.0	nc	Mixed	TULE-12	na	<1	38.6	nc	Pre-Modern
KING-37	21	5.4	0.0	nc	Modern	TULE-13	17	4.4	0.0	nc	Modern
KING-38	21	<1	82.9	nc	Pre-Modern	TULE-14	18	2.1	0.0	nc	Modern
KING-39	17	<1	82.4	nc	Pre-Modern	TULE-15	nc	nc	nc	nc	Not Datable
KWH-01	17	7.0	7.9	nc	Mixed	TULE-16	21	2.5	0.0	nc	Modern
KWH-02	17	4.5	0.0	nc	Modern	TULE-17	19	<1	1.0	nc	Mixed
KWH-03	19	8.9	0.0	nc	Modern						

Table D5.   Groundwater age classification information for USGS-grid and USGS-understanding wells sampled October 2005 through February 2006, Southeast San Joaquin Valley study unit, California GAMA Priority Basin Project.—Continued

[Samples classified as pre-modern if recharged berfore 1952. Samples classified as modern if recharged after 1952. Samples classified as mixed if sample contains both modern and pre-modern water. **Abbreviations**: USGS, U.S. Geological Survey; GAMA, Groundwater Ambient Monitoring and Assessment Program; °C, degrees Celsius; TU, tritium units; KING, Kings study area well; KWH, Kaweah study area well; TLR, Tulare Lake study area well; TULE, Tule study area well; FP, flow-path well; HWY99T, transect well; na, not available; <, less than; nc, not collected]

USGS GAMA well identification number	Noble-gas based recharge temperature (°C)	Tritium activity (TU)	Terrigenic helium, (percent of total helium)	Modern carbon-14 (percent)	Groundwater age classification	USGS GAMA well identification number	Noble-gas based recharge temperature (°C)	Tritium activity (TU)	Terrigenic helium, (percent of total helium)	Modern carbon-14 (percent)	Groundwater age classification
Understanding wells						**Understanding wells—Continued**					
KINGFP-01	20	1.6	33.0	105	Mixed	KINGFP-09	22	<1	33.3	93.0	Pre-Modern
KINGFP-02	22	<1	85.5	47.9	Pre-Modern	KINGFP-10	19	<1	75.8	nc	Pre-Modern
KINGFP-03	21	<1	80.8	42.8	Pre-Modern	KINGFP-11	21	2.2	9.4	121	Mixed
KINGFP-04	20	2.4	5.9	102	Mixed	KINGFP-12	20	2.6	0.0	124	Modern
KINGFP-05	20	3.2	11.0	nc	Mixed	KINGFP-13	20	1.7	0.0	91.6	Modern
KINGFP-06	20	<1	78.3	79.2	Pre-Modern	KINGFP-14	29	3.6	0.0	97.3	Modern
KINGFP-07	26	3.9	0.0	114	Modern	KINGFP-15	19	3.9	0.0	96.7	Modern
KINGFP-08	27	6.8	0.0	118	Modern	HWY99T-01	16	5.1	27.6	nc	Mixed

Table D6. Groundwater age classification information for USGS-grid and USGS-understanding wells sampled January through March 2006, Kern County Subbasin study unit, California GAMA Priority Basin Project.

[Samples classified as pre-modern if recharged before 1953. Samples classified as modern if recharged after 1953. Samples classified as mixed if sample contains both modern and pre-modern water. **Abbreviations**: USGS, U.S. Geological Survey; GAMA, Groundwater Ambient Monitoring and Assessment Program; °C, degrees Celsius; TU, tritium units; KERN, Kern County Subbasin well; FP, flow-path well; nc, not collected; na, not available; <, less than]

USGS GAMA well identification number	Noble-gas based recharge temperature (°C)	Tritium activity (TU)	Terrigenic helium, (percent of total helium)	Modern carbon-14 (percent)	Groundwater age classification
Grid wells					
KERN-01	18.9	2.6	0.0	nc	Modern
KERN-02	nc	nc	nc	nc	nc
KERN-03	19.3	1.3	66.6	nc	Mixed
KERN-04	16.4	<1	0.0	nc	Mixed
KERN-05	13.7	<1	86.3	nc	Pre-Modern
KERN-06	18.3	<1	14.9	nc	Pre-Modern
KERN-07	22.1	<1	0.0	nc	Mixed
KERN-08	19.5	<1	90.7	nc	Pre-Modern
KERN-09	16.8	4.0	0.0	nc	Modern
KERN-10	na	<1	73.8	nc	Pre-Modern
KERN-11	21.4	<1	0.0	nc	Mixed
KERN-12	19.4	<1	36.5	nc	Pre-Modern
KERN-13	17.4	<1	0.0	nc	Mixed
KERN-14	15.3	<1	75.9	nc	Pre-Modern
KERN-15	16.7	2.9	0.0	nc	Modern
KERN-16	19.3	3.7	22.9	nc	Mixed
KERN-17	17.9	6.5	0.0	nc	Modern
KERN-18	15.5	7.1	0.0	nc	Modern
KERN-19	na	1.0	0.0	nc	Modern
KERN-20	nc	nc	nc	nc	nc
KERN-21	17.2	<1	0.0	nc	Mixed
KERN-22	16.7	<1	16.0	nc	Pre-Modern
KERN-23	21.8	1.5	44.0	nc	Mixed
KERN-24	23.5	2.9	0.0	nc	Modern
KERN-25	13.9	3.3	11.0	nc	Mixed
KERN-26	16.8	1.4	2.7	nc	Modern
KERN-27	nc	nc	nc	nc	nc
KERN-28	19.8	<1	27.6	nc	Pre-Modern
KERN-29	20.3	8.0	0.0	nc	Modern
KERN-30	22.0	<1	0.9	nc	Mixed
KERN-31	20.6	5.0	0.0	nc	Modern
KERN-32	17.2	<1	26.6	nc	Pre-Modern
KERN-33	16.5	<1	45.4	nc	Pre-Modern
KERN-34	17.0	2.7	0.0	90.3	Modern
KERN-35	19.9	3.4	0.0	104	Modern
KERN-36	17.4	6.7	20.8	62.9	Mixed
KERN-37	55.8	3.4	0.0	90.0	Mixed
KERN-38	22.2	1.0	0.0	71.3	Mixed
KERN-39	19.9	4.7	0.0	98.4	Modern
KERN-40	16.9	<1	38.3	nc	Pre-Modern
KERN-41	16.3	<1	66.3	1.75	Pre-Modern
KERN-42	15.7	<1	0.0	83.2	Mixed

Table D6.   Groundwater age classification information for USGS-grid and USGS-understanding wells sampled January through March 2006, Kern County Subbasin study unit, California GAMA Priority Basin Project.—Continued

[Samples classified as pre-modern if recharged before 1953. Samples classified as modern if recharged after 1953. Samples classified as mixed if sample contains both modern and pre-modern water. **Abbreviations**: USGS, U.S. Geological Survey; GAMA, Groundwater Ambient Monitoring and Assessment Program; °C, degrees Celsius; TU, tritium units; KERN, Kern County Subbasin well; FP, flow-path well; nc, not collected; na, not available; <, less than]

USGS GAMA well identification number	Noble-gas based recharge temperature (°C)	Tritium activity (TU)	Terrigenic helium, (percent of total helium)	Modern carbon-14 (percent)	Groundwater age classification
Grid wells—Continued					
KERN-43	19.5	<1	48.9	nc	Pre-Modern
KERN-44	19.6	1.5	0.0	89.4	Mixed
KERN-45	19.6	<1	18.6	64.0	Pre-Modern
KERN-46	18.0	<1	1.7	nc	Mixed
KERN-47	nc	nc	nc	nc	nc
Understanding wells					
KERNFP-01	17.9	7.7	0.0	nc	Modern
KERNFP-02	17.4	9.2	0.0	90.5	Modern
KERNFP-03	17.4	4.8	0.0	nc	Modern

Table D7. Oxidation-reduction classification and pH for USGS-grid and USGS-understanding wells, Southeast San Joaquin Valley study unit, California GAMA Priority Basin Project.

[Redox category and redox process determined using the algorithm of McMahon and Chapelle (2008) implemented by Jurgens and others (2009b) except for samples with incomplete redox data, which were excluded from the analysis. **Abbreviations**: USGS, U.S. Geological Survey; GAMA, Groundwater Ambient Monitoring and Assessment Program; mg/L, milligram per liter; µg/L, microgram per liter; redox, oxidation-reduction; KING, Kings study area well; KWH, Kaweah study area well; TLR, Tulare Lake study area well; TULE, Tule study area well; FP, flow-path well; HWY99T, transect well; nc, not collected; –, not detected; <, less than; ≥, greater than or equal to; anoxic-NO$_3$, nitrate reducing; oxic, dissolved oxygen ≥ 0.5; suboxic, dissolved oxygen < 0.5; anoxic-Mn, manganese reducing; anoxic-Fe, iron reducing; O$_2$, oxygen; Fe(III), iron oxide; SO$_4$, sulfate; Mn(IV), manganese oxide]

USGS GAMA well identification number	pH	Oxidizing and reducing constituents					Redox category	Redox process
		Dissolved oxygen (mg/L)	Nitrate plus nitrite (mg/L)	Manganese (µg/L)	Iron (µg/L)	Sulfate (mg/L)		
Grid wells								
KING-01	nc	0.1	–	nc	nc	nc	O$_2$ < 0.5 mg/L	Unknown
KING-02	nc	0.2	–	nc	nc	nc	O$_2$ < 0.5 mg/L	Unknown
KING-03	8.7	0.1	–	–	180	19.0	Anoxic	Fe(III)
KING-04	8.2	2	2.53	0.3	5	22.4	Oxic	O$_2$
KING-05	nc	0.1	–	nc	nc	nc	O$_2$ < 0.5 mg/L	Unknown
KING-06	nc	0.1	nc	nc	nc	nc	O$_2$ < 0.5 mg/L	Unknown
KING-07	8.1	1.1	0.68	–	–	14.0	Oxic	O$_2$
KING-08	7.9	4.4	0.90	–	–	5.0	Oxic	O$_2$
KING-09	8.2	4.1	1.79	0.1	–	8.5	Oxic	O$_2$
KING-10	7.8	5.3	1.69	–	–	8.8	Oxic	O$_2$
KING-11	7.9	5.7	4.09	1.5	7	7.6	Oxic	O$_2$
KING-12	7.9	7.9	8.24	0.1	–	34.6	Oxic	O$_2$
KING-13	8	6.9	3.21	0.1	–	5.3	Oxic	O$_2$
KING-14	8.7	3.2	2.03	–	–	4.0	Oxic	O$_2$
KING-15	7.9	6.5	6.95	0.2	–	7.5	Oxic	O$_2$
KING-16	7.7	8.7	2.09	0.2	–	4.1	Oxic	O$_2$
KING-17	7.9	4.7	4.80	0.1	–	4.3	Oxic	O$_2$
KING-18	7.2	6.6	0.84	–	–	4.8	Oxic	O$_2$
KING-19	7.7	4.3	3.57	–	–	10.3	Oxic	O$_2$
KING-20	7.4	4.5	1.01	–	3	7.9	Oxic	O$_2$
KING-21	nc	0.3	–	nc	nc	nc	O$_2$ < 0.5 mg/L	Unknown
KING-22	8.0	6.6	0.90	–	–	0.0	Oxic	O$_2$
KING-23	7.9	9.8	3.39	–	–	14.0	Oxic	O$_2$
KING-24	7.6	5.9	5.33	–	–	27.0	Oxic	O$_2$
KING-25	7.5	3.1	5.14	0.3	4	13.3	Oxic	O$_2$
KING-26	nc	0.3	nc	nc	nc	nc	O$_2$ < 0.5 mg/L	Unknown
KING-27	nc	6.1	3.25	nc	nc	nc	O$_2$ ≥ 0.5 mg/L	Unknown
KING-28	nc	2.2	0.56	nc	nc	nc	O$_2$ ≥ 0.5 mg/L	Unknown
KING-29	7.0	3.10	1.49	–	–	4.5	Oxic	O$_2$
KING-30	7.1	5.1	2.75	–	–	7.0	Oxic	O$_2$
KING-31	nc	1.3	nc	nc	nc	nc	O$_2$ ≥ 0.5 mg/L	Unknown
KING-32	nc	1.0	–	nc	nc	nc	O$_2$ ≥ 0.5 mg/L	Unknown
KING-33	nc	1.9	nc	nc	nc	nc	O$_2$ ≥ 0.5 mg/L	Unknown
KING-34	nc	4.6	5.35	nc	nc	nc	O$_2$ ≥ 0.5 mg/L	Unknown
KING-35	nc	6.7	3.68	nc	nc	nc	O$_2$ ≥ 0.5 mg/L	Unknown
KING-36	8.1	4.2	1.58	nc	nc	nc	O$_2$ ≥ 0.5 mg/L	Unknown
KING-37	nc	7.0	nc	nc	nc	nc	O$_2$ ≥ 0.5 mg/L	Unknown
KING-38	7.9	3.2	2.45	0.2	6	6.1	Oxic	O$_2$
KING-39	nc	0.1	nc	nc	nc	nc	O$_2$ < 0.5 mg/L	Unknown
KWH-01	8.0	3.1	4.47	–	–	17.0	Oxic	O$_2$
KWH-02	7.8	6.3	1.36	–	–	5.0	Oxic	O$_2$
KWH-03	8.1	6.7	2.65	–	4	13.1	Oxic	O$_2$
KWH-04	8.0	3.7	1.81	–	–	9.0	Oxic	O$_2$
KWH-05	nc	7.0	21.69	nc	nc	nc	O$_2$ ≥ 0.5 mg/L	Unknown

Table D7.   Oxidation-reduction classification and pH for USGS-grid and USGS-understanding wells, Southeast San Joaquin Valley study unit, California GAMA Priority Basin Project.—Continued

[Redox category and redox process determined using the algorithm of McMahon and Chapelle (2008) implemented by Jurgens and others (2009b) except for samples with incomplete redox data, which were excluded from the analysis. **Abbreviations**: USGS, U.S. Geological Survey; GAMA, Groundwater Ambient Monitoring and Assessment Program; mg/L, milligram per liter; µg/L, microgram per liter; redox, oxidation-reduction; KING, Kings study area well; KWH, Kaweah study area well; TLR, Tulare Lake study area well; TULE, Tule study area well; FP, flow-path well; HWY99T, transect well; nc, not collected; –, not detected; <, less than; ≥, greater than or equal to; anoxic-NO_3, nitrate reducing; oxic, dissolved oxygen ≥ 0.5; suboxic, dissolved oxygen < 0.5; anoxic-Mn, manganese reducing; anoxic-Fe, iron reducing; O_2, oxygen; Fe(III), iron oxide; SO_4, sulfate; Mn(IV), manganese oxide]

USGS GAMA well identification number	pH	Oxidizing and reducing constituents					Redox category	Redox process
		Dissolved oxygen (mg/L)	Nitrate plus nitrite (mg/L)	Manganese (µg/L)	Iron (µg/L)	Sulfate (mg/L)		
Grid wells—Continued								
KWH-06	8.1	8.8	9.13	0.5	22	28.3	Oxic	O_2
KWH-07	nc	2.1	10.84	nc	nc	nc	$O_2 \geq 0.5$ mg/L	Unknown
KWH-08	nc	7.9	nc	nc	nc	nc	$O_2 \geq 0.5$ mg/L	Unknown
KWH-09	nc	8.3	5.65	nc	nc	nc	$O_2 \geq 0.5$ mg/L	Unknown
KWH-10	8.2	11	14.46	–	–	37.0	Oxic	O_2
KWH-11	9.0	0.4	–	2.3	15	6.7	Suboxic	Suboxic
KWH-12	8.5	5.8	3.26	0.4	9	12.9	Oxic	O_2
KWH-13	nc	3.9	7.12	nc	nc	nc	$O_2 \geq 0.5$ mg/L	Unknown
KWH-14	7.9	7.9	2.71	–	–	10.5	Oxic	O_2
KWH-15	nc	4.1	–	nc	nc	nc	$O_2 \geq 0.5$ mg/L	Unknown
KWH-16	nc	9.5	nc	nc	nc	nc	$O_2 \geq 0.5$ mg/L	Unknown
KWH-17	nc	5.2	13.10	nc	nc	nc	$O_2 \geq 0.5$ mg/L	Unknown
KWH-18	nc	7.8	nc	nc	nc	nc	$O_2 \geq 0.5$ mg/L	Unknown
TLR-01	8.2	10.9	2.71	–	120	69.0	Mixed (oxic-anoxic)	O_2-Fe(III)/SO_4
TLR-02	8.8	1.0	–	2.3	20	0.3	Oxic	O_2
TLR-03	8.4	0.1	–	–	45.4	15.7	Suboxic	Suboxic
TLR-04	8.8	0.1	–	16.9	6	0.2	Suboxic	Suboxic
TLR-05	9.4	0.2	–	1	26	1.2	Suboxic	Suboxic
TLR-06	8.3	0.9	–	66.2	181	282.0	Mixed (oxic-anoxic)	O_2-Fe(III)/SO_4
TLR-07	7.8	0.2	–	28	–	228.0	Suboxic	Suboxic
TLR-08	8.6	1.8	–	6.6	5	13.8	Oxic	O_2
TLR-09	nc	0.1	nc	nc	nc	nc	$O_2 < 0.5$ mg/L	Unknown
TULE-01	8.8	4.0	3.37	–	6	18.3	Oxic	O_2
TULE-02	7.9	7.3	5.42	–	–	23.0	Oxic	O_2
TULE-03	8.6	1.5	2.30	0.1	4	13.5	Oxic	O_2
TULE-04	8.2	0.6	5.74	20	–	70.0	Oxic	O_2
TULE-05	nc	0.4	1.13	nc	nc	nc	$O_2 < 0.5$ mg/L	Unknown
TULE-06	nc	0.9	nc	nc	nc	nc	Indeterminate	
TULE-07	8.3	5.5	2.34	–	–	12.0	Oxic	O_2
TULE-08	9.8	1.1	0.43	0.2	6	8.5	Oxic	O_2
TULE-09	7.3	1.0	1.15	–	–	8.0	Oxic	O_2
TULE-10	nc	7.5	6.78	nc	nc	nc	$O_2 \geq 0.5$ mg/L	Unknown
TULE-11	nc	0.3	nc	nc	nc	nc	$O_2 < 0.5$ mg/L	Unknown
TULE-12	nc	0.1	nc	nc	nc	nc	$O_2 < 0.5$ mg/L	Unknown
TULE-13	nc	2.2	3.16	nc	nc	nc	$O_2 \geq 0.5$ mg/L	Unknown
TULE-14	7.6	8.6	7.68	–	–	48.0	Oxic	O_2
TULE-15	nc	0.4	nc	nc	nc	nc	$O_2 < 0.5$ mg/L	Unknown
TULE-16	nc	6.9	nc	nc	nc	nc	$O_2 \geq 0.5$ mg/L	Unknown
TULE-17	8.0	6.8	1.75	–	–	13.2	Oxic	O_2
Understanding wells								
KINGFP-01	7.4	6.2	12.1	–	9	64.2	Oxic	O_2
KINGFP-02	8.0	3.5	2.82	–	5	3.6	Oxic	O_2
KINGFP-03	8.1	3.5	3.54	–	6	3.6	Oxic	O_2

Table D7. Oxidation-reduction classification and pH for USGS-grid and USGS-understanding wells, Southeast San Joaquin Valley study unit, California GAMA Priority Basin Project.—Continued

[Redox category and redox process determined using the algorithm of McMahon and Chapelle (2008) implemented by Jurgens and others (2009b) except for samples with incomplete redox data, which were excluded from the analysis. **Abbreviations**: USGS, U.S. Geological Survey; GAMA, Groundwater Ambient Monitoring and Assessment Program; mg/L, milligram per liter; µg/L, microgram per liter; redox, oxidation-reduction; KING, Kings study area well; KWH, Kaweah study area well; TLR, Tulare Lake study area well; TULE, Tule study area well; FP, flow-path well; HWY99T, transect well; nc, not collected; –, not detected; <, less than; ≥, greater than or equal to; anoxic-NO_3, nitrate reducing; oxic, dissolved oxygen ≥ 0.5; suboxic, dissolved oxygen < 0.5; anoxic-Mn, manganese reducing; anoxic-Fe, iron reducing; O_2, oxygen; Fe(III), iron oxide; SO_4, sulfate; Mn(IV), manganese oxide]

USGS GAMA well identification number	pH	Oxidizing and reducing constituents					Redox category	Redox process
		Dissolved oxygen (mg/L)	Nitrate plus nitrite (mg/L)	Manganese (µg/L)	Iron (µg/L)	Sulfate (mg/L)		
Understanding wells—Continued								
KINGFP-04	7.5	6.5	6.41	0.1	–	23.5	Oxic	O_2
KINGFP-05	7.1	4.5	7.17	0.5	11	64.8	Oxic	O_2
KINGFP-06	7.7	5.7	2.38	0.1	5	5.9	Oxic	O_2
KINGFP-07	6.6	6.1	25.8	0.2	<6	112.0	Oxic	O_2
KINGFP-08	7.2	8.3	12.8	28.4	4	98.5	Oxic	O_2
KINGFP-09	7.4	9.1	2.80	0.4	7	11.1	Oxic	O_2
KINGFP-10	7.6	5.3	2.44	–	–	13.1	Oxic	O_2
KINGFP-11	7.5	6.4	10.3	0.8	5	124.0	Oxic	O_2
KINGFP-12	7.2	6.1	34.3	0.2	–	360.0	Oxic	O_2
KINGFP-13	7.5	4.9	10.3	–	–	27.0	Oxic	O_2
KINGFP-14	7.3	0.1	–	2,910	50	14.4	Anoxic	Mn(IV)
KINGFP-15	7.3	0.1	–	651	168	9.2	Anoxic	SO_4
HWY99T-01	7.8	6.8	1.76	–	–	8.0	Oxic	O_2

Table D8. Oxidation-reduction classification and pH for USGS-grid and USGS-understanding wells, Kern County Subbasin study unit, California GAMA Priority Basin Project.

[Redox category and redox process determined using the algorithm of McMahon and Chapelle (2008) implemented by Jurgens and others (2009b) except for samples with incomplete redox data, which were excluded from the analysis. **Abbreviations**: USGS, U.S. Geological Survey; GAMA, Groundwater Ambient Monitoring and Assessment Program; mg/L, milligram per liter; µg/L, microgram per liter; redox, oxidation-reduction; KERN, Kern County Subbasin well; FP, flow-path well; nc, not collected; <, less than; –, not detected; ≥, greater than or equal to; oxic, dissolved oxygen ≥ 0.5; suboxic, dissolved oxygen < 0.5; anoxic-NO$_3$, nitrate reducing; anoxic-Mn, manganese reducing; anoxic-Fe, iron reducing; O$_2$, oxygen; Fe(III), iron oxide; SO$_4$, sulfate; NO$_3$, nitrate; Mn(IV), manganese oxide]

USGS GAMA well identification number	pH	Oxidizing and reducing constituents					Redox category	Redox process
		Dissolved oxygen (mg/L)	Nitrate plus nitrite (mg/L)	Manganese (µg/L)	Iron (µg/L)	Sulfate (mg/L)		
Grid wells								
KERN-01	nc	7.5	nc	nc	nc	nc	O$_2$ ≥ 0.5 mg/L	Unknown
KERN-02	nc	3.9	nc	nc	nc	nc	O$_2$ ≥ 0.5 mg/L	Unknown
KERN-03	8.2	1.9	6.10	–	–	100	Oxic	O$_2$
KERN-04	8.1	1.1	1.04	–	–	20	Oxic	O$_2$
KERN-05	nc	1.8	nc	nc	nc	nc	O$_2$ ≥ 0.5 mg/L	Unknown
KERN-06	nc	3.9	4.29	–	–		O$_2$ ≥ 0.5 mg/L	Unknown
KERN-07	8.2	1.4	8.81	–	–	63	Oxic	O$_2$
KERN-08	nc	<0.2	–	nc	nc	nc	Suboxic	Unknown
KERN-09	6.8	2.1	1.94	–	–	nc	O$_2$ ≥ 0.5 mg/L	Unknown
KERN-10	8.13	<0.2	<0.10	12	230	554	Anoxic	Fe(III)/SO$_4$
KERN-11	nc	5.6	4.29	nc	nc	nc	O$_2$ ≥ 0.5 mg/L	Unknown
KERN-12	nc	1.8	1.54	nc	nc	nc	O$_2$ ≥ 0.5 mg/L	Unknown
KERN-13	8.6	7.7	3.16	–	–	62	Oxic	O$_2$
KERN-14	8.3	<0.2	–	58	1,800	2.3	Anoxic	Fe(III)
KERN-15	7.8	1.7	2.60	10	100	26	Mixed (oxic-anoxic)	O$_2$-SO$_4$
KERN-16	nc	0.3	nc	nc	nc	nc	Suboxic	Unknown
KERN-17	8.0	5.4	0.38	–	–	22	Oxic	O$_2$
KERN-18	8.2	6.6	0.69	0.19	18.8	11.4	Oxic	O$_2$
KERN-19	9.3	<0.2	0.01	–	–	64	Suboxic	Suboxic
KERN-20	nc	6.4	nc	nc	nc	nc	O$_2$ ≥ 0.5 mg/L	Unknown
KERN-21	nc	5.7	nc	nc	nc	nc	O$_2$ ≥ 0.5 mg/L	Unknown
KERN-22	8.5	1.1	3.73	–	–	91	Oxic	O$_2$
KERN-23	nc	0.5	–	nc	nc	nc	O$_2$ ≥ 0.5 mg/L	Unknown
KERN-24	nc	3.6	nc	nc	nc	nc	O$_2$ ≥ 0.5 mg/L	Unknown
KERN-25	nc	1.9	nc	nc	nc	nc	O$_2$ ≥ 0.5 mg/L	Unknown
KERN-26	7.7	0.8	–	–	–	44	Mixed (oxic-anoxic)	O$_2$-Fe(III)-SO$_4$
KERN-27	nc	6.9	nc	nc	nc	nc	O$_2$ ≥ 0.5 mg/L	Unknown
KERN-28	nc	0.2	nc	nc	nc	nc	Suboxic	Unknown
KERN-29	nc	5.0	–	nc	nc	nc	O$_2$ ≥ 0.5 mg/L	Unknown
KERN-30	nc	8.4	nc	nc	nc	nc	O$_2$ ≥ 0.5 mg/L	Unknown
KERN-31	nc	4.9	nc	nc	nc	nc	O$_2$ ≥ 0.5 mg/L	Unknown
KERN-32	8.1	<0.2	–	22	60	426	Suboxic	Suboxic
KERN-33	nc	1.1	nc	nc	nc	nc	O$_2$ ≥ 0.5 mg/L	Unknown
KERN-34	8.0	7.8	1.92	–	–	34	Oxic	O$_2$
KERN-35	8.1	5.8	1.04	–	–	23.7	Oxic	O$_2$
KERN-36	8.5	3.0	1.12	0.2	–	38.8	Oxic	O$_2$
KERN-37	7.7	2.4	0.36	2.6	8	28.3	Oxic	O$_2$
KERN-38	7.7	4.0	13.4	0.7	–	131	Oxic	O$_2$
KERN-39	7.9	7.8	12.4	0.3	–	120	Oxic	O$_2$
KERN-40	7.8	0.2	2.78	6.3	22	779	Anoxic	NO$_3$
KERN-41	9.6	0.5	3.38	0.5	13	39.7	Oxic	O$_2$
KERN-42	8.4	1.5	0.55	0.5	–	12.1	Oxic	O$_2$
KERN-43	7.7	<0.2	0.54	1.2	–	43.4	Anoxic	NO$_3$
KERN-44	8.0	9.0	8.50	–	–	30.3	Oxic	O$_2$

Table D8. Oxidation-reduction classification and pH for USGS-grid and USGS-understanding wells, Kern County Subbasin study unit, California GAMA Priority Basin Project.—Continued

[Redox category and redox process determined using the algorithm of McMahon and Chapelle (2008) implemented by Jurgens and others (2009b) except for samples with incomplete redox data, which were excluded from the analysis. **Abbreviations**: USGS, U.S. Geological Survey; GAMA, Groundwater Ambient Monitoring and Assessment Program; mg/L, milligram per liter; µg/L, microgram per liter; redox, oxidation-reduction; KERN, Kern County Subbasin well; FP, flow-path well; nc, not collected; <, less than; –, not detected; ≥, greater than or equal to; oxic, dissolved oxygen ≥ 0.5; suboxic, dissolved oxygen < 0.5; anoxic-NO$_3$, nitrate reducing; anoxic-Mn, manganese reducing; anoxic-Fe, iron reducing; O$_2$, oxygen; Fe(III), iron oxide; SO$_4$, sulfate; NO$_3$, nitrate; Mn(IV), manganese oxide]

USGS GAMA well identification number	pH	Dissolved oxygen (mg/L)	Nitrate plus nitrite (mg/L)	Manganese (µg/L)	Iron (µg/L)	Sulfate (mg/L)	Redox category	Redox process
Grid wells—Continued								
KERN-45	9.1	3.0	2.04	–	13	95.5	Oxic	O$_2$
KERN-46	7.6	<0.2	7.45	69.3	–	579	Mixed (anoxic)	NO$_3$-Mn(IV)
KERN-47	8.7	5.4	4.48	0.4	10	161	Oxic	O$_2$
Understanding wells								
KERNFP-01	8.0	<0.2	–	80.3	31	147	Anoxic	Mn(IV)
KERNFP-02	7.9	<0.2	0.07	1.3	7	21.1	Suboxic	Suboxic
KERNFP-03	7.1	4.0	0.46	6.9	–	20.3	Oxic	O$_2$

Table D9.   Oxidation-reduction classification and pH for domestic wells sampled in Tulare County and located in the Southeast San Joaquin Valley study unit, California GAMA Domestic Well Project.

[Redox classification was not determined because of incomplete data. **Abbreviations**: USGS, U.S. Geological Survey; GAMA, Groundwater Ambient Monitoring and Assessment Program; mg/L, milligram per liter; μg/L, microgram per liter; TUL, Tulare County well; nc, not collected; –, not detected]

USGS GAMA well identification number	pH	Oxidizing and reducing constituents					USGS GAMA well identification number	pH	Oxidizing and reducing constituents				
		Dissolved oxygen (mg/L)	Nitrate plus nitrite (mg/L)	Manganese (μg/L)	Iron (μg/L)	Sulfate (mg/L)			Dissolved oxygen (mg/L)	Nitrate plus nitrite (mg/L)	Manganese (μg/L)	Iron (μg/L)	Sulfate (mg/L)
TUL901	6.9	nc	nc	1.72	87.5	nc	TUL952	7.1	nc	2.8	–	–	18
TUL902	6.5	nc	3.4	1.86	42.4	15	TUL955	7.5	nc	9.9	9.28	–	41
TUL903	6.5	nc	–	10.2	56.4	6.3	TUL958	7.0	nc	50	2.49	–	160
TUL904	7.1	nc	21	2.94	115	120	TUL959	7.0	nc	11	7.33	–	80
TUL905	6.8	nc	1.6	5.24	–	13	TUL960	7.4	nc	9.2	0.57	–	40
TUL906	7.0	nc	3.6	–	27.1	33	TUL961	7.3	nc	12	1.66	–	34
TUL907	6.8	nc	14	1.81	–	69	TUL967	7.3	nc	8.1	3.13	–	32
TUL908	6.8	nc	–	2.91	107	2.5	TUL968	6.5	nc	4.8	1.36	–	18
TUL909	6.3	nc	5.2	2.65	46.8	11	TUL969	7.0	nc	6.8	172	–	60
TUL910	6.8	nc	6.8	25.8	–	150	TUL970	7.1	nc	16	–	21.1	49
TUL911	6.9	nc	0.61	5.14	168	5.7	TUL971	7.0	nc	13	2.15	39	33
TUL912	6.6	nc	9.5	–	–	77	TUL972	7.3	nc	6.5	1.38	–	28
TUL913	6.3	nc	0.75	–	–	3.5	TUL973	7.2	nc	44	3.43	–	93
TUL914	7.0	nc	7	–	–	67	TUL974	6.7	nc	21	7.95	162	160
TUL915	6.3	nc	4.7	1.79	–	6.4	TUL976	7.3	nc	3	1.87	–	9.3
TUL916	6.7	nc	7	–	–	31	TUL978	7.1	nc	21.9	32.8	–	81
TUL917	6.6	nc	8.4	–	–	50	TUL979	7.4	nc	54	0.71	–	52
TUL922	7.1	nc	14	–	–	71	TUL986	7.6	nc	5.1	0.83	35.3	16
TUL923	6.6	nc	1.4	1.33	–	9.4	TUL987	7.5	nc	2.7	1.82	–	28
TUL924	6.8	nc	9	–	23.7	49	TUL988	7.2	nc	18.3	2.7	–	104
TUL925	6.9	nc	4.7	10.4	–	100	TUL990	7.4	nc	6.3	2.39	–	48
TUL926	6.9	nc	–	1.26	–	14	TUL991	7.9	nc	8.4	8.96	189	17
TUL928	7.6	nc	1.6	1.8	83.9	35	TUL992	7.2	nc	12	1.85	–	28
TUL929	6.8	nc	25	–	–	50	TUL996	7.6	nc	5.4	1.15	–	16
TUL930	7.2	nc	21	4.95	68.6	47	TUL997	7.4	nc	27	2.78	–	46
TUL931	7.2	nc	13	1.84	–	60	TUL999	7.2	nc	7.5	2.89	–	43
TUL932	7.1	nc	25	2.6	153	120	TUL1000	7.5	nc	8.4	3.33	–	16
TUL933	7.2	nc	9.8	3.51	235	79	TUL1003	7.7	nc	2.5	1.37	–	5.5
TUL934	7.6	nc	18	–	219	48	TUL1005	7.8	nc	9.26	1.39	–	31.6
TUL935	7.2	nc	14	0.92	53.4	50	TUL1006	7.4	nc	7.81	2.91	–	19.9
TUL936	7.4	nc	1.5	–	–	3.3	TUL1007	7.5	nc	6.05	1.76	–	18.9
TUL937	8.4	nc	7	–	–	10	TUL1008	6.9	nc	21.8	0.7	–	122
TUL938	8.4	nc	4.3	–	–	14	TUL1010	7.5	nc	0.11	0.72	–	5.7
TUL939	7.6	nc	5.2	–	–	15	TUL1011	7.4	nc	14	7.44	650	84
TUL940	7.1	nc	0.15	–	–	3.5	TUL1012	8.1	nc	3.8	0.75	–	16
TUL941	7.4	nc	19	4.68	–	14	TUL1013	6.9	nc	41	1.52	–	53
TUL942	7.0	nc	9.2	–	–	23	TUL1014	7.1	nc	32	0.88	–	70
TUL943	6.9	nc	6.3	–	–	23	TUL1015	6.9	nc	35	2.03	–	110
TUL944	7.3	nc	11	–	–	19	TUL1016	7.2	nc	11	2.75	–	180
TUL945	7.5	nc	16	–	–	43	TUL1017	7.5	nc	11	1.38	–	88
TUL946	7.1	nc	8.3	–	–	14	TUL1020	7.1	nc	31	0.55	–	42
TUL947	7.1	nc	4.3	–	68.6	39	TUL1021	6.9	nc	31	1.43	–	78
TUL948	7.9	nc	1.3	–	–	14	TUL1022	7.4	nc	–	30.1	–	110
TUL949	7.3	nc	8	–	–	66	TUL1025	7.0	nc	28	2.52	–	220
TUL950	7.7	nc	7.3	–	81	32	TUL1026	7.0	nc	17	1.04	–	88
TUL951	7.5	nc	3.2	–	–	44	TUL1027	7.7	nc	0.42	0.87	–	2.5

Table D9. Oxidation-reduction classification and pH for domestic wells sampled in Tulare County and located in the Southeast San Joaquin Valley study unit, California GAMA Domestic Well Project.—Continued

[Redox classification was not determined because of incomplete data. **Abbreviations**: USGS, U.S. Geological Survey; GAMA, Groundwater Ambient Monitoring and Assessment Program; mg/L, milligram per liter; μg/L, microgram per liter; TUL, Tulare County well; nc, not collected; –, not detected]

USGS GAMA well identification number	pH	Dissolved oxygen (mg/L)	Nitrate plus nitrite (mg/L)	Manganese (μg/L)	Iron (μg/L)	Sulfate (mg/L)	USGS GAMA well identification number	pH	Dissolved oxygen (mg/L)	Nitrate plus nitrite (mg/L)	Manganese (μg/L)	Iron (μg/L)	Sulfate (mg/L)
TUL1028	7.4	nc	7.4	1.68	–	18	TUL1072	6.7	nc	4.79	2.78	–	42.3
TUL1029	7.4	nc	4.9	4.39	125	18	TUL1073	7.2	nc	12.1	3.11	–	44.7
TUL1031	7.7	nc	–	2.64	–	3.8	TUL1074	7.1	nc	11	2.24	–	62.9
TUL1032	7.1	nc	16	2.99	49.9	64	TUL1075	7.1	nc	10.1	0.55	–	62.2
TUL1033	7.6	nc	7.2	1.13	20.1	30	TUL1076	6.9	nc	10.9	0.75	–	96.4
TUL1034	7.1	nc	0.34	3.52	–	7.5	TUL1080	7.5	nc	10.1	0.63	–	33.9
TUL1035	7.8	nc	6.64	1.62	85.4	55.6	TUL1081	7.0	nc	50.8	0.54	–	45.6
TUL1036	6.6	nc	0.66	0.87	–	4.7	TUL1082	7.7	nc	3.86	0.99	–	6.5
TUL1041	7.2	nc	14	0.86	–	35	TUL1083	7.3	nc	5.33	2.28	–	20
TUL1042	6.8	nc	0.43	3.56	42.9	3.9	TUL1084	7.3	nc	3.48	16.4	–	44
TUL1043	7.1	nc	9.64	2.34	–	55.4	TUL1085	6.7	nc	31.8	1.3	–	58.6
TUL1044	7.4	nc	2.21	2.15	–	13.5	TUL1088	7.0	nc	4.22	0.82	–	30.3
TUL1054	8.1	nc	4.97	2.2	30.5	29.2	TUL1089	7.1	nc	22.4	18.3	50.7	87.7
TUL1055	7.4	nc	0.72	2.06	–	3.5	TUL1091	7.0	nc	11	0.88	–	34
TUL1056	6.8	nc	1.2	15.1	50.8	ND	TUL1092	7.2	nc	15.7	1.19	–	77.6
TUL1057	6.8	nc	21.5	1.22	–	64.9	TUL1093	7.4	nc	10.5	8.32	–	24
TUL1058	6.8	nc	28.9	11.6	100	36.7	TUL1094	6.9	nc	2.05	–	–	ND
TUL1059	7.2	nc	20.4	0.88	–	26.9	TUL1098	7.9	nc	0.56	2.88	–	5.2
TUL1060	7.1	nc	4.97	3.69	–	50.8	TUL1101	7.3	nc	43.8	0.11	–	28.4
TUL1061	6.7	nc	10	1.34	–	99.5	TUL1103	7.6	nc	2.51	19.7	–	7.1
TUL1062	7.2	nc	20.7	0.56	101	35.3	TUL1105	7.3	nc	11.1	0.33	–	78.6
TUL1064	7.6	nc	1.49	0.54	–	5.8	TUL1106	8.1	nc	2.01	9.43	–	6.1
TUL1065	6.8	nc	13.1	19.2	–	37.8	TUL1107	7.2	nc	18.7	0.13	–	64.6
TUL1066	7.5	nc	0.56	11.5	57.2	2.5	TUL1111	7.0	nc	50.4	13.4	–	84.3
TUL1070	7.0	nc	24.6	2.36	–	64.8	TUL1505	7.3	nc	20.5	2.9	–	54.5
TUL1071	6.6	nc	6.82	8.1	–	47.2